.

Biogenic Amines

Edited by Charalampos Proestos

Published in London, United Kingdom

IntechOpen

Supporting open minds since 2005

Biogenic Amines
http://dx.doi.org/10.5772/intechopen.75221
Edited by Charalampos Proestos

Contributors
Georgios Danezis, Aristidis S. Tsagkaris, Antonios-Dionysios G. Neofotistos, Charalampos Proestos, Maria Martuscelli, Dino Mastrocola, Dincer Erdag, Oguz Merhan, Baris Yildiz, M. Carmen Vidal-Carou, Oriol Comas-Basté, Mariluz Latorre-Moratalla, Sònia Sánchez-Pérez, M. Teresa Veciana-Nogués

Notice
Statements and opinions expressed in the chapters are these of the individual contributors and not necessarily those of the editors or publisher. No responsibility is accepted for the accuracy of information contained in the published chapters. The publisher assumes no responsibility for any damage or injury to persons or property arising out of the use of any materials, instructions, methods or ideas contained in the book.

First published in London, United Kingdom, 2019 by IntechOpen
IntechOpen is the global imprint of INTECHOPEN LIMITED, registered in England and Wales, registration number: 11086078, The Shard, 25th floor, 32 London Bridge Street
London, SE19SG – United Kingdom
Printed in Croatia

British Library Cataloguing-in-Publication Data
A catalogue record for this book is available from the British Library

Additional hard copies can be obtained from orders@intechopen.com

Biogenic Amines
Edited by Charalampos Proestos
p. cm.
Print ISBN 978-1-78984-133-6
Online ISBN 978-1-78984-134-3

Meet the editor

Charalampos Proestos has a BSc (Ptychio) in Chemistry from the University of Ioannina, Greece, and an MSc in Food Science from Reading University, UK. He obtained his PhD in Food Chemistry at the Agricultural University of Athens (AUA), Greece, where he continued his postdoc working on natural antioxidants on programs funded by the European Union and Greece. After further training at Wageningen University (Netherlands), he worked as a research associate at AUA. He also worked as a chemist for the Hellenic Food Authority, being the food industry auditor and supervisor of the Chemical Laboratory in Athens accredited with ISO 17025. Currently, he is an assistant professor at the Department of Chemistry, National and Kapodistrian University of Athens. He has published more than 60 papers in reputed journals and has been serving as an editorial board member of more than 10 reputable journals. He is a member of the European Committee of the Division of Food Chemistry, European Association of Chemical and Molecular Sciences (EuChemS). His research field focuses on food antioxidants, foodomics, and food contaminants.

Contents

Preface

Human health can be affected by minor food compounds that sometimes are neglected by consumers. Biogenic amines (BAs) are naturally occurring amines and are a group belonging to the aforementioned compounds. Their concentrations are positively affected by microbial fermentation of food or during food spoilage. Many analytical methods are published regarding extraction and characterization of these minor food substances.

BAs are nitrogen-containing compounds formed by the decarboxylation of amino acids or by amination and transamination of ketones and aldehydes. The structure of BAs can be aliphatic, aromatic, and heterocyclic. Depending on the number of amine groups, amines are classified into monoamines (tyramine and phenylethylamine), diamines (histamine, putrescine, and cadaverine), or polyamines (spermidine and spermine). Currently, there are research papers that have classified cadaverine, putrescine, spermine, and spermidine among polyamines.

Proestos Charalampos
Assistant Professor in Food Chemistry,
Department of Chemistry, Laboratory of Food Chemistry,
National and Kapodistrian University of Athens,
Athens, Greece

Section 1

Introduction

Chapter 1

Introductory Chapter: Current Knowledge on Biogenic Amines

Proestos Charalampos

1. Introduction

Human health can be affected by minor food compounds which sometimes are neglected by consumers. Biogenic amines are naturally occurred amines and for sure they can be one group belonging to the aforementioned compounds. Their concentrations are positively affected by microbial fermentation of food or during food spoilage. Many analytical methods are published regarding extraction and characterization of these minor food substances. Biogenic amines (BAs) are nitrogen-contained compounds occurring by decarboxylation of amino acids or by amination and ransamination of ketones and aldehydes. The structure of BAs can be aliphatic, aromatic, and heterocyclic. Depending on the number of amine groups, amines are classified into monoamines (tyramine and phenylethylamine), diamines (histamine, putrescine, and cadaverine), or polyamines (spermidine and spermine). Currently, there are research papers that have classified cadaverine, putrescine, spermine, and spermidine among polyamines [1].

2. BAs occurrence in foods and health effect

BAs are found in a variety of food products including seafood, meat, dairy, fruits, vegetables, nuts, chocolates, and fermented products. Foods containing proteins or amino acids in free form can have microbial or biochemical activity and BAs may be present.

Fermented products can contain BAs such as milk products, for example, cheese 5–4500 ppm, wines 5–135 ppm, beer 2.7–14 ppm. BAs can also be present in foods that are spoiled or in the beginning of spoilage such as fish and fish products 2300–5000 ppm, animal liver like beef about 330 ppm, processed meat and meat products 9–650 ppm [2]. BAs can be used in nonfermented products such as decomposition markers or indicator compounds. Deteriorated foods contain high amounts of BAs with the most abundant being cadaverine and putrescine. BAs contained in high amounts can have a serious health risk for consumers who belong to sensitive population. Symptoms may include hot flushes, respiratory discomfort, nausea, cold sweat, headaches, palpitations, red rash, low or high blood pressure. Alcohol consuming and presence of acetaldehyde may increase sensitivity to biogenic amines.

3. Safety issues and function

Scombroid poisoning which is caused due to histamine presence is a very important issue that is why only histamine has a regulatory limit, according to EU regulation (maximum 200 mg/kg in fresh fish and 400 mg/kg in fish products after enzyme treatment and maturation in brine). After fish, cheese is the next most commonly implicated food item associated with tyramine poisoning, so called "cheese reaction," related with its high content in aged cheeses. Other potential BAs, especially histamine and putrescine, are also present in milk-based fermented foods. Among the approaches useful to control the formation of BAs are the reduction of microbial growth through chilling and freezing or hydrostatic pressures, irradiation, controlled atmosphere packaging, or the use of food additives, etc.

Proteins and protein-like compounds such as alkaloids, hormones, and nucleic acids can be synthesized by precursors such BAs which contain nitrogen. BAs could probably affect organism in processes like body temperature regulation, nutrient intake, blood pressure decrease or increase. As far as plants are concerned, putrescine (diamine) and spermidine and spermine (polyamines) play important role in physiological processes, such as flowering, cell division, fruit development, and stress response. Spermine and spermidine are essential for growth, metabolism, and renovation, of almost all organs in the human body and important for high metabolic activity maintenance of the proper functioning and immunological gut system [3].

BAs are potential precursors for the formation of carcinogenic N-nitroso compounds. In lipid-containing foods, such as ham, bacon, at high temperature and in high water content, the carcinogenic compound N-nitrosopyrrolidine can be produced from spermidine or putrescine. BAs like putrescine, cadaverine, spermidine may play the role of free radical quenchers. Tyramine is regarded as a very good antioxidant, and its antioxidant capacity increases with its concentration. Free radical inhibition depends on amino and hydroxy groups. Spermine, for example, can regenerate tocopherol from tocopheroxyl radical through hydrogenic donor from amino group.

The radical formed binds lipid or peroxide radicals into lipid complexes. Some BAs contribute to the flavor and taste of food.

4. BAs analytical methods

The analytical methods reported for identification and quantification of BAs are based on different types of chromatography: gas chromatography (GC), thin layer chromatography (TLC), and high-performance liquid chromatography (HPLC) combined with precolumn or postcolumn derivatization techniques. Aliphatic BAs do not have absorption bands in the UV-VIS region, hence simple spectrometric detectors cannot be applied [4, 5]. Analysis of BAs without prior derivatization includes ion-pair chromatography combined with octylamine or heptanesulfonate as ion-pair reagents. For BAs ion-pair separation in reversed-phase columns with C12-C18 aliphatic chains, phenyl residues bound to a silica core are efficient. HPLC procedures usually involve precolumn or postcolumn derivatization procedure. Chemical reagents that are usually used for BA analysis by postcolumn derivatization are mainly ninhydrin and o-phthalaldehyde and for precolumn derivatization the following reagents: dansyl and dabsyl chloride, fluorescein, benzoyl chloride, 9-fluorenylmethyl chloroformate [5].

5. Conclusion

BAs are low-molecular-mass organic bases which occur in plant- and animal-derived products. BAs in food can occur by free amino acid enzymatic decarboxylation and other metabolic processes. Usually, in the human body, amines contained in foods are quickly detoxified by enzymes such as amine oxidases or by conjugation; however, in allergic individuals or if monoamino oxidase (MAO) inhibitors are applied, the detoxification process is disturbed and BAs accumulate in the body. Knowing the concentration of BAs is essential because they can affect human health, and also because they can be used as freshness indicators to estimate the degree of food spoilage.

Author details

Proestos Charalampos
Department of Chemistry, Laboratory of Food Chemistry, National and Kapodistrian University of Athens, Athens, Greece

*Address all correspondence to: harpro@chem.uoa.gr

IntechOpen

References

[1] Koutsoumanis K, Tassou C, Nychas G. Biogenic amines in foods. In: Juneja V, Sofos J, editors. Pathogens and Toxins in Foods. Washington, DC: ASM Press; 2010. pp. 248-274. DOI: 10.1128/9781555815936.ch16

[2] EFSA Panel on Biological Hazards (BIOHAZ). Scientific opinion on risk based control of biogenic amine formation in fermented foods. EFSA Journal. 2011;**9**:2393

[3] Linares DM, Río BD, Ladero V, Martínez N, Fernández M, Martín MC, et al. Factors influencing biogenic amines accumulation in dairy products. Frontiers in Microbiology. 2012;**3**:180. DOI: 10.3389/fmicb.2012.00180

[4] Erim FB. Recent analytical approaches to the analysis of biogenic amines in food samples. Trends in Analytical Chemistry. 2013;**52**:239-247

[5] Proestos C, Loukatos P, Komaitis M. Determination of biogenic amines in wines by HPLC with precolumn dansylation and fluorimetric detection. Food Chemistry. 2008;**106**(3):1218-1224

Trends in Biogenic Amines Analysis

Chapter 2

Emerging Trends in Biogenic Amines Analysis

Antonios-Dionysios G. Neofotistos, Aristeidis S. Tsagkaris,
Georgios P. Danezis and Charalampos Proestos

Abstract

Biogenic amines are low-molecular-mass substances, essential for proper health for all organisms. These compounds could be detrimental to human health with various toxicological effects when they are present in high concentrations. Therefore, biogenic amines monitoring in food samples is a matter of utmost importance, and their accurate determination is considered indispensable. Under this context, we provide an overview over the most widely employed analytical techniques for biogenic amines determination such as chromatographic techniques and biosensors, emphasizing on new approaches. A critical comparison of the techniques is also given, presenting their advantages and drawbacks regarding important analytical characteristics such as sensitivity. Finally, we focus on foods in which biogenic amines mainly occur such as fish, meat and wine and other fermented products.

Keywords: biogenic amines, analysis, food, chromatography, biosensors

1. Introduction

Biogenic amines (BAs) are small molecular organic nitrogenous compounds (bases), polar or semipolar. The most common BAs in food are putrescine, cadaverine, spermine and spermidine with an aliphatic structure; tyramine, tryptamine and β-phenylethylamine with an aromatic structure and histamine with a heterocyclic structure [1, 2]. These compounds can produce a wide range of toxicological effects [3]. Theoretically, BAs occurrence could be expected in all foods that contain free amino acids (AAs) or protein and are exposed to conditions enabling microbial or biochemical activity [4]. Biogenic amines share common characteristics with their precursors, AAs, and that is taken into consideration when we try to come up with analytical methods for their effective determination.

In low concentrations, these nitrogenous organic bases are essential for good health, acting as hormones or neurotransmitters and generally being important for growth, temperature regulation and high metabolic activity of the normal functioning and immunological system of gut. By and large, BAs constitute sources of nitrogen and precursors which lead to the synthesis of many specific compounds such as hormones, alkaloids, proteins and nucleic acids. Also, they control several processes in the organism such as the regulation of body temperature, intake of nutrition and increase/decrease of blood pressure [5].

On the other hand, in high concentrations, BAs are considered quite hazardous and able to cause health problems to consumers, especially to sensitive persons.

Histamine is the most widely studied biogenic amine due to its ability to cause headaches, nausea, hypotension, digestive problems and skin allergy, while tyramine is often associated with migraine and hypertension [6]. It has also been proved that tyramine is more cytotoxic than histamine on an in vitro model of the human intestinal epithelium. Tyramine caused a cell necrosis, while histamine induced apoptosis [7]. Referring to polyamines such as putrescine and cadaverine, they are less pharmacologically active; however, they could interact with the amine oxidases and potentiate the effects of histamine and tyramine. Besides, these polyamines can react with nitrite to form potentially carcinogenic nitrosamines [8]. At this point, it should be noted that microbiological spoilage cannot occur in salted products since BAs accumulation can occur before salting. Moreover, if sea salt, rock salt or other preservatives contain nitrate and nitrite as impurities, BAs in salted products may react with nitrites to form nitrosamines, as mentioned above [9, 10]. Tyramine, cadaverine, putrescine and histamine are the most common BAs in meat and meat products. The concentration of histamine is usually quantitatively lower than that found in fish [11].

As for the BAs level regulations, histamine is currently the only BAs having official limits in fish products, despite the fact that BAs have been described as having a certain potential toxicity in food products in general. The European Food Safety Authority confirmed histamine and tyramine as the most toxic and particularly relevant for food safety [11]. The presence of other amines has been found to enhance histamine toxicity [9, 10]. The maximum acceptable histamine levels in fish have been established in many countries; in the USA, the Food and Drug Administration established a maximum limit of 50 mg kg^{-1} at the port and 100 mg kg^{-1} in pickled fish for species prone to form histamine [11, 12]. The European Union has established regulations according to which histamine levels should be below 100 mg kg^{-1} in raw fish and below 200 mg kg^{-1} in salted fish [13], regarding species belonging in the Coryphaenidae, Engraulidae, Pomatomidae, Clupeidae and Scombridae families. In Brazil, the Regulation of Industrial and Sanitary Inspection of Animal Products does not mention the amine maximum level allowed in products of animal origin. However, the MERCOSUR institute, co-run by Argentina, Brazil, Paraguay and Uruguay, established a maximum level of 100 mg kg^{-1} of histamine in the muscles of species of the Clupeidae, Pomatomidae, Scombridae, Scomberesocidae and Coryphaenidae families [14]. Scombroid poisoning is a type of fish poisoning developed by eating spoiled fish. Since most fish species are rich in free histidine, scombroid poisoning is accepted as mainly caused by elevated histamine levels in fish, generated by bacterial enzymatic conversion of free histidine [9]. Fish products head the list of foods most studied from the point of view of BAs.

Some European countries have recommended the establishment limits for histamine in wine: Germany (2 mg L^{-1}), Belgium (5–6 mg L^{-1}) and France (8 mg L^{-1}) [15, 16]. Switzerland established an upper limit of 10 mg L^{-1} of histamine, which was rejected later [17]. In Slovak Republic, histamine is regulated as 20 mg kg^{-1} in beer, 200 mg kg^{-1} in fish/fish products and tyramine as 200 mg kg^{-1} in cheese [18]. The Institute of Dairy Research in the Netherlands and the Czech Republic has proposed an upper limit of 100–200 mg kg^{-1} for histamine in meat products. There are no official establishments regarding the standards for cadaverine, putrescine or other BAs, except for some proposals. About tyramine, the recommended limit is in the range of 100–800 mg kg^{-1} of food. A figure of 30 mg kg^{-1} for β-phenylethylamine has been considered toxic dose in food [1].

Since the occurrence of BAs has been detected in a broad range of products (fish, fish products, meat, meat products, beer, wine, cheese, milk, dairy products, various beverages, condiments, fruits, vegetables, vinegar, tea, chocolate and coffee), it is an urgent need to develop new analytical methods or improve the current methods for BAs analysis in terms of rapidity and reliability as far as food safety is concerned.

The aim of this chapter is to summarize, discuss and compare the most widely employed analytical techniques/approaches such as chromatographic, capillary electrophoresis and biosensors for BAs. Moreover, we provide the emerging trends and the recent advances on BAs analytical methods.

2. Techniques

2.1 Chromatographic

The determination of biogenic amines is not simple at all, due to the variety of their chemical structures and their presence at relatively low levels in matrices which are usually complex. However, a precise identification of these compounds and the detection of even slight changes in their profile are urgent regarding both the quality control and the consumers' health. The monitoring and determination of BAs in food matrices is based on new analytical approaches which combine higher accuracy, sensitivity, reproducibility and rapidity and are also inexpensive and convenient and thus can be adopted by laboratories worldwide, applied in numerous applications. In this part, the most widely employed analytical techniques are described and discussed. Moreover, newly applied methods are presented, compared and evaluated, based on innovations and technical possibilities that they propose.

The quantification of BAs in food samples has mainly been accomplished by an array of chromatographic methods, such as high-performance liquid chromatography (HPLC), gas chromatography (GC), ultra-performance liquid chromatography (UPLC), ion chromatography (IC), thin-layer chromatography (TLC), ion-pair liquid chromatography (IPLC) and capillary electrophoresis (CE). However, in recent years, many sensors have been developed for the BAs analysis as alternatives to the expensive instrumentation of chromatographic techniques [19]. What is more, it should be noted that the employment of a sensitive and efficient detector is a crucial issue as it ensures the trustworthiness of the analytical method as a whole. There are several detection processes having been reported in BAs detection studies, such as ultraviolet (UV) [20], indirect UV [21], mass spectrometry (MS) [22], electrochemical [23], conductometric [24], enzymatic, immunoassay and polymerase chain reaction (PCR) processes [25]. Generally, after derivatization, the approaches usually used for BAs determination are UV, fluorescence and MS.

Classical reversed-phase high-performance liquid chromatography (RP-HPLC) using C-18 columns has been indistinctively employed for the BAs quantitative determination in different types of food because of its sensitivity, high resolution, great versatility and relatively simple sample treatment [26]. Yet, a previous solvent extraction step and a chemical derivatization step is required prior to final separation. The former aims to remove some potentially interfering compounds and also to concentrate the analytes of interest. The latter reduces the BAs polarity and improves resolution in RP columns, making them more sensitive towards detection. Also, the BAs polarity needs to be reduced because this high polar character results in a greater solubility in water rather than in the organic solvents which are frequently used in the majority of the techniques.

Solid phase extraction (SPE) is a widely used technique for sample clean-up and proper isolation of BAs, while it is the most common extraction method for BAs determination in beverages [27]. Also, solid phase microextraction (SPME) [28], molecularly imprinted solid phase extraction (MISPE) [29], dispersive liquid-liquid microextraction (DLLME) [30], vortex-assisted surfactant-enhanced emulsification liquid-liquid microextraction (VSLLME) [31], hollow-fibre liquid-phase microextraction (HF-LPME) [32], salting-out assisted liquid-liquid extraction (SALLE) [33] and cloud point extraction (CPE) [34] have been used in many different BAs determination studies. Also, the addition of specific chemical substances is sometimes required so as to ensure the retention of potentially interfering substances such as lipids, proteins and polyphenols. These compounds have similar structures to BAs, thereby posing problems for the derivatization reaction, making the BAs quantification and detection difficult. The chemical substances added usually are trichloroacetic acid (TCA), ethyl acetate, hydrochloric acid, perchloric acid, diethyl ether and polyvinylpyrrolidone (PVP). Hence, the matrix interferences are minimized, up to an extent.

The selection of an effective derivatization agent is a matter of utmost importance in order to decrease the derivatization time and increase the derivatization reaction efficiency. There are many different HPLC studies discussed, aiming to BAs determination and using numerous derivatization agents such as 6-aminoquinolyl-N-hydroxysuc cinimidyl carbamate (AQC) [35], dansyl chloride (Dns-Cl) [36–39], O-phthalaldehyde (OPA) [19], 2,6-dimethyl-4-quinolinecarboxylic acid N-hydroxysuccinimide ester (DMQC-Osu) [40], ethyl-acridine-sulfonyl chloride (EAC) [41], O-phthalaldehyde/N-acetyl-L-cysteine (OPA/NAC) [42], O-phthalaldehyde/mercaptoethanol (OPA/MCE) [43] or 1,3,5,7-tetramethyl-8-(N-hydroxysuccinimidyl butyric ester)-difluorobora-diaza-s-indacene (TMBBSu) [44] coupled with fluorescence detection. It is obvious that the obtained limits of detection (LODs) were quite satisfying. Also, there are some HPLC studies presented, employing benzoyl chloride [45], 2-chloro-1,3-dinitro-5-(trifluoromethyl)-benzene (CNBF) [46, 47], diethyl ethoxymethylenemalonate (DEEMM) [48], 9-fluorenylmethyl chloroformate (FMOC) [31], 1-naphthylisothiocyanate (NITC) [49], phenyl isothiocyanate (PITC) [50] or dabsyl chloride (Dbs-Cl) [51] with UV detection, obtaining low LODs, too.

By and large, the LODs values of analytical methods for BAs usually lie at ppm (mg L^{-1}) levels. In some cases, SPE processes prior to chromatographic analysis can decrease the LODs values even to ppb (μg L^{-1}) levels, and the ultra-trace analysis of BAs can be accomplished. For instance, in a study of Basheer et al. [28], the synthesized hydrazone-based ligands were physically trapped in a silica sol-gel matrix and used for micro-SPE of the dansylated BAs. The technique was applied to the pre-concentration of BAs in orange juice, before HPLC analysis and UV detection. The sol-gel sorbent that contained benzophenone 2,4-dinitrophenylhydrazone ligand showed great affinity to the target analytes. The obtained LODs were 3.82–31.3 ng L^{-1}. Apart from that, the LODs of 0.25–50 μg L^{-1} were obtained by Huang et al. [40], applying the IL-based ultrasonic-assisted liquid-liquid microextraction method (IL-UALLME) with DMQC-OSu as derivatization agent. Also the LODs of 8.82–40.4 ng L^{-1} were obtained in a study by Gao et al. with TMBBSu as a derivatization agent [44].

All in all, the methods using a fluorescent detector are more sensitive than those with a UV-vis detector, regardless of the reagent employed. In two studies in Chilean young and reserved wine samples, conducted by the same laboratory and using the same protocol, the LODs using a fluorescence detector were 1–90 μg L^{-1}, whereas the LODs using a UV detector were 90–300 μg L^{-1} [36, 37]. In the same way, in a HPLC study in fish products, the LODs with fluorescence detection were 20–240 μg kg^{-1}, while the LODs with UV detection were 567–1800 μg kg^{-1} [38, 52]. However, the study of Tameem et al. [53] using a UV-vis detector and obtaining sensibly low LODs (20–60 ng L^{-1}) represents an exception to the above mentioned. The authors stated that this accomplishment was because of the large sample volume (50 mL) injected into the chromatograph.

In some cases, the derivatization step can be phased out when HPLC is coupled with MS detectors. Hence, the analysis time can be much smaller. Sagratini et al.

elaborated a method with underivatized BAs in fish tissues using LC-MS/MS analysis after SPE [54]. The obtained LODs were 20–250 µg kg^{-1}. On the whole, due to the considerable sensitivity and the specific structural information for the derivatized amines that MS or MS/MS detectors can provide, they are the most efficient detection tools for metabolites which are usually present in low concentrations [55]. Furthermore, they are very helpful in identifying co-eluting peaks in real sample analysis. What is more, tandem mass spectrometry (MS/MS) was employed in a study where isotopically labeled BAs were added as internal standards. Isotopically labeled internal standards have been proven to minimize matrix interferences in complex matrices. The limits of quantification (LOQs) were at 50 ng kg^{-1} level [56]. Finally, in a very recent study, Jastrzębska et al. (2018) used 3,5-Bis-(trifluoromethyl)phenyl isothiocyanate (BPI) to produce the BAs-BPI derivatives which were determined by liquid chromatography-tandem mass spectrometry (LC-MS/MS). The obtained LODs ranged from 2.0 to 4.3 ng L^{-1} [57].

In another method, the separation of BAs with HPLC was followed by evaporative light-scattering detection (ELSD) [58]. The detector's response was based on the amount of light scattered by analyte particles created by evaporating a solvent while it passed through a light beam. There was no need for the use of chromophores for target analytes; hence no derivatization was required. This LC-ELSD method was compared to a LC-UV method applied in the same study and was found less sensitive. Yet, it was good enough for the BAs detection in cheese samples. The LODs were 1.4–3.6 mg L^{-1}. In general, ELS detectors are considered more affordable than mass spectrometers with the same characteristics.

In a BAs determination study in beer samples, dairy beverage samples and herb tea and vinegar samples, the analytes were separated by ion-pair liquid chromatography and detected by a chemiluminescent nitrogen detector (CLND) [59]. In comparison with a HPLC-UV and a HPLC-charged aerosol detector (CAD) method for 14 BAs in the same study, the HPLC-CLND method gave narrower peaks, with highly improved resolutions. The LODs were 0.1–0.4 mg L^{-1}. Sun et al. [59] optimized and validated this method using nonafluoropentanoic acid (NFPA), which is an ideal agent as it was tested and selected as the finest ion-pair reagent amongst many other perfluorocarboxylic acids tried. In another implementation of ion-pair liquid chromatography (IPLC), the chaotropic salt KPF$_6$ was applied in vinegar samples [60].

It should be noted that the time of analysis depends heavily on the number of analytes and usually varies between 5 and 85 min when the conventional LC method with C18 columns is employed. In the vast majority of studies with simultaneous determination of BAs, the run time lasted more than 30 min. In many cases lately, the BAs determination has been performed by UPLC, which outweighs HPLC, mainly in terms of solvent consumption, analysis time, better resolution and increased peak capacities [61]. Since 2004, new generations of stationary phases compatible with LC systems have been widely used under the trade name Ultra Performance Liquid Chromatography (UPLC) [62]. The separation time of BAs is greatly decreased when the UPLC technique with short columns (5 cm) packed with smaller particles (<2 µm) and high flow rates is employed [19]. Jia et al. [63] published a study where they developed an ultra-performance LC/quadrupole time-of-flight mass spectrometry method (UPLC/Q-TOFMS) for dansylated BAs along with 23 amino acids in cheese, beer and sausage samples. The separation of all analytes was completed in 25–30 min. The LODs were 5–20 µg L^{-1}. The UPLC/Q-TOFMS method was also used for BAs detection in another study [64] with less analysis time than the conventional HPLC method (13 min) and LODs 3–15 µg L^{-1}. In another UPLC-MS/MS method, the elution time was also very short (<8.5 min) [65]. In a more recent study by Lee et al. [66], the UPLC-MS/MS (ESI+) method was applied for the determination of 9 BAs in rice wine samples with run time 21 min and LODs 0.1–4.6 µg L^{-1}.

Next, in some cases, BAs have been determined by ion chromatography (IC). The LODs were not very low, yet this technique does not require a derivatization step. The LODs in fruit juice samples were 56–1630 µg L^{-1} [67], whereas the LODs in wine samples were 23–68 µg kg^{-1} [68]. Ion chromatography (IC) coupled with pulsed amperometric detection (PAD) or integrated pulsed amperometric detection (IPAD) has been reported in many studies. All the same, the use of acids and salts in conjunction with the use of organic solvents in high concentrations is sometimes essential for the separation of strongly retained amines, such as spermidine and spermine [69]. Organic solvents can cause decomposition by-products resulting in potential interferences. Hence, longer retention times and poor resolution or peak shapes have been reported [70].

Thin-layer chromatography (TLC) exemplifies an alternative to LC methods. No special equipment is required, and several samples can be analysed at the same time. Notwithstanding, this method is semi-quantitative and the analysis can be relatively long [71, 72]. An economic TLC/densitometry for BAs detection in wine samples was validated [71]. The potential interferences were avoided with the use of PVPP. Furthermore, isohexane was used for the extraction of the dansylated derivatives, before TLC separation. The analysis was completed in 90 min and the LODs were 600–700 µg L^{-1}.

Gas chromatography (GC) methods are also used in some studies, yet not to the same extent that LC or capillary electrophoresis (CE) does, due to the lack of volatility of BAs. Apart from decreasing the polarity of BAs, the derivatization step in GC is essential so as to increase the volatile properties of these analytes. Mass spectrometers are used for detection in the majority of studies. Almeida et al. validated a dispersive liquid-liquid microextraction (DLLME) method followed by GC-MS for the determination of 18 BAs in beer samples. The DLLME procedure was performed simultaneously with the derivatization process. The LODs were 0.3–2.9 µg L^{-1} [73]. Also, Cunha et al. elaborated a GC-MS method and used IBCF to determine the content of 22 BAs in grape juice and wine samples, with toluene as an extraction solvent and LODs lying at the level of 1 µg L^{-1}. The derivatization was carried out in a two-phase reaction system, eliminating the need for a previous extraction procedure [74]. The IBCF derivatization was also performed in a BAs determination study in home-made fermented alcoholic drinks by Plotka-Wasylka et al. [75] involving in situ derivatization-DLLME combined with GC-MS. The LODs were 1.1–4.1 µg L^{-1} [75]. In a more recent study, Huang et al. performed an environmentally friendly SPME coupled with GC-MS for the determination of biogenic amines in fish samples. The LODs were 2.98–45.3 µg kg^{-1} [76]. The analysis time was shorter in the GC methods than the LC methods, and the obtained LODs were generally satisfying.

2.2 Capillary electrophoresis

Capillary electrophoresis (CE) is a separation technique which is widely performed, following HPLC regarding the extent of its application in biogenic amine analysis. This technique renders possible the analysis of a vast range of compounds which occur in low concentration levels and is also characterized by rapidity, separation efficiency, sensitivity and shorter analysis time than LC methods and less solvent consumption. Another important fact is that CE is suitable for analytes which cannot be analysed with GC due to thermal instability. Yet, it should be noted that the number of separated BAs and their precursors, AAs, is usually much smaller than HPLC methods [69]. There are different CE methods employed in BAs studies, such as micellar electrokinetic chromatography (MEKC), capillary gel electrophoresis (CGE), capillary zone electrophoresis (CZE), capillary isotachophoresis (CITP) and capillary isoelectric focusing (CIEF) [77].

Fluorometric methods are frequently employed by virtue of the fluorescence of BAs at some pH range their reaction with proper agents. Many fluorescence derivatization reagents have been used in CE studies, like fluorescein isothiocyanate (FITC), 5-(4,6-dichloro-s-triazin-2-ylamino) fluorescein (DTAF), naphthalene-2,3-dicarboxaldehyde (NDA), OPA and 3-(2-furoyl)-quinoline-2-carboxaldehyde (FQ). In a study involving FITC, proposed by Uzaşçı et al., a fast separation of seven biogenic amines in wine samples was performed, and significantly low LODs down to 57.6–113 ng L^{-1} were reported [78]. The authors proposed a novel nonionic micellar electrokinetic chromatography method (MEKC) through FITC coupled to laser-induced fluorescence detector (LIF). The separation was completed in 9 min. Also, in a study by Zhang et al. involving the combination of laser-induced fluorescence (LIF) detector and CE separation, a quite satisfying improvement in detection limits was observed [79].

A CE-MS/MS method was reported for the quantitative determination of BAs in beer and wine samples [80]. The migration time for the 9 BAs was very short (<10 min), and the LODs were in the range of 1–2 μg L^{-1} for wine and 3–8 μg L^{-1} for beer samples. The main drawback regarding the use of a MS detector for CE is its higher price compared with conventional UV or LIF detection and the limitation in the type of running buffers that can be used as they have to be volatile and compatible with ESI.

Offline precolumn derivatization is the most widely used application, while the occurring BAs derivatives are then injected into the CE. Generally, UV detection sensitivities in CE are lower than those of HPLC and can be increased by novel online pre-concentration procedures along with derivatization developments or coupling CE with isotachophoresis (ITP) [19].

In a study, the online coupling of capillary zone electrophoresis (CZE) with capillary isotachophoresis (CITP) and UV detection increased the sensitivity of the method. The BAs were online pre-concentrated in the ITP step, separated and detected in the CZE step. The study was conducted for the determination of histamine, phenylethylamine and tyramine in wine samples. The LODs were 0.35 mg L^{-1} for histamine, 0.33 mg L^{-1} for phenylethylamine and 0.37 mg L^{-1} for tyramine [81]. In a more recent CITP method with a conductometric detector in beer and wine samples by Jastrzębska et al. [82], the derivatization step was eliminated, and the LODs 200–480 μg L^{-1} were lower in comparison with an LC method by the same authors. In the latter, dansylated BAs derivatives were synthesized and a UV detector was employed.

Finally, in a method elaborated by Dossi et al., the separation of BAs and AAs in beer samples was performed by microchip CE and followed by amperometric detection with the use of ruthenium oxide/hexacyanoruthenate polymeric films, electrochemically deposited onto glassy carbon electrodes. The separation of single amines and AAs was not possible through this method. Hence, the analytes were co-eluted in groups. The LODs were 1.4–6.8 mg L^{-1} [83].

2.3 Biosensors

A biosensor is widely considered as an analytical platform that converts a biological response into a quantifiable and processable signal. The individual parts of a biosensor were well addressed and comprehensively discussed in [84], which is recommended for further understanding of biosensors function. Even though instrumental methods are quite accurate and sensitive, their high operational cost and time-consuming protocols showcase the need for analytical alternatives. In this way, biosensors can improve the current situation as they feature highly desired characteristic such as simplicity, rapidness, cost-effectiveness and portability. In BAs analysis, chromatographic techniques are mostly used as the numerous compounds of food matrices need to be separated to accurately detect the analytes [26]. However, biosensors, instead of separation, utilize various selective recognition elements such as antibodies, enzymes, molecularly imprinted polymers (MIPs), aptamers and nucleic acids to bind with the target molecule.

As it is already shown, the utilisation of biosensors in BAs analysis is a novel concept that finds more and more applications in food analysis. Under this context, we indicatively discuss some published studies to provide an overview on the various recognition elements and detection methods. To begin with, various sensing elements have been used (**Table 1**), including classic approaches such as antibodies or enzymes and also pioneering cases such MIPs or nanotechnological applications. Despite the excellent recognition capability of MIPs or the proven superior electronic properties of nanocomposite materials, antibodies and enzymes are still mostly used because of the laborious development process of those recognition elements. Concerning the detection method, electrochemical (EC) sensors were commonly used as they did not require time-consuming sample pretreatment and could be easily miniaturized and used in situ. However, EC sensors usually lacked long-term stability. Alternatively, spectroscopic methods, namely chemiluminescence (CL) and photoluminescence (PL), were also applied because they combined simple formats with zero background measurements for increased sensitivity. Interestingly, digital cameras were also used for the colorimetric detection of BAs. The combination of image data analysis with dipsticks [85] or immunoassays [86] is a trend focusing towards portable and on-site detection of various food contaminants. A striking example of this is the EU-funded FoodSmartphone project, http://foodsmartphone.eu/ (last visited 5/8/2018), in which several smartphone-based assays for the detection of various food contaminants including allergens,

Analyte	Recognition element	Detection	Matrix	LODs	Ref.
Tyramine	Antibody	ELISA	Meat and fish	1.2 mg kg^{-1}	[87]
Putrescine	Unsaturated complex of Cu(II)	CL	Shrimp	0.0178 mg L^{-1}	[88]
Histamine	Cu@Pd core-shell nanostructures	EC	Tuna fish	0.3 ng kg^{-1}	[89]
Tryptamine	Oxidation	EC	Banana, tomato cheese, sausages	0.12 ng L^{-1}	[90]
Tyramine	Fe^{3+} ion complex of FONs	Ratiometric PL	Solutions	0.5 mg L^{-1}	[91]
Tyramine	Dye Py-1	Digital camera	Shrimp	1.37 mg kg^{-1}	[92]
Tyramine	Tyrosinase	EC	Cheese, sauerkraut, banana	0.21 mg L^{-1}	[93]
Histamine	Antibody	EC	Fish	1.25 ng L^{-1}	[94]
Tryptamine	Covalent immobilization of tryptamine on nanofiber	PL	Beer	6 ng L^{-1}	[95]
Tyramine	MIPs	EC	Milk	0.32 ng L^{-1}	[96]

Table 1.
Biosensors application in BAs analysis.

pesticides, marine toxins etc. are being developed. Hence, this approach may also find application for the BAs in the future. Regarding biosensors sensitivity, we can notice that they provided satisfactory detection limits, but in several studies there were no validation data in the food matrix. All in all, biosensors have the potential to improve and simplify the current situation and move towards the on-site BAs determination.

3. Conclusions

The determination of BAs in food is a matter of utmost importance regarding both the consumers' health and the quality control. Regulatory bodies, industries and consumers demand efficient detection of BAs in food products. Thus, the scientific community tries to develop new analytical methods or to improve the current methods concerning their sensitivity and reliability in different food matrices. Yet, it should be noted that the legislation for BAs regulations may differ amongst different countries; there are different production, processing and storage methods, as well as different climate conditions which can either cause or inhibit the BAs formation in food [97, 98]. Classical reversed-phase high-performance liquid chromatography (RP-HPLC) using C-18 columns has been the most commonly employed technique for the BAs quantitative determination because of its sensitivity, high resolution, great versatility and relatively simple sample treatment. However, CE exemplifies a good alternative to HPLC, with high separation efficiency and relatively low running costs. Lastly, the development on sensors has led to the elaboration of new methods characterized by low cost, short analysis time and simplicity. Neither special instrumentation nor sample clean-up and derivatization are required.

Acknowledgements

G.P. Danezis was funded through a postdoctoral studies scholarship, from the Hellenic State Scholarship Foundation (IKY) that is gratefully acknowledged. This research is supported through IKY scholarships programme and co-financed by the European Union (European Social Fund—ESF) and Greek national funds through the action entitled "Reinforcement of Postdoctoral Researchers", in the framework of the Operational Programme "Human Resources Development Program, Education and Lifelong Learning" of the National Strategic Reference Framework (NSRF) 2014–2020.

Author details

Antonios-Dionysios G. Neofotistos[1], Aristeidis S. Tsagkaris[2], Georgios P. Danezis[1*]
and Charalampos Proestos[3]

1 Chemistry Laboratory, Department of Food Science and Human Nutrition,
Agricultural University of Athens, Athens, Greece

2 Department of Food Analysis and Nutrition, Faculty of Food and Biochemical
Technology, University of Chemistry and Technology, Prague, Czech Republic

3 Department of Chemistry, Food Chemistry Laboratory, National and
Kapodistrian University of Athens, Athens, Greece

*Address all correspondence to: gdanezis@aua.gr

IntechOpen

References

[1] Papageorgiou M et al. Literature update of analytical methods for biogenic amines determination in food and beverages. TrAC Trends in Analytical Chemistry. 2018;**98**:128-142

[2] Shalaby AR. Significance of biogenic amines to food safety and human health. Food Research International. 1996;**29**(7):675-690

[3] E. P. o. B. Hazards. Scientific opinion on risk based control of biogenic amine formation in fermented foods. EFSA Journal. 2011;**9**(10):2393

[4] Halász A et al. Biogenic amines and their production by microorganisms in food. Trends in Food Science & Technology. 1994;**5**(2):42-49

[5] Karovičová J, Kohajdová Z. Biogenic amines in food. Chemical Papers. 2005;**59**(1):70-79

[6] Ladero V et al. Toxicological effects of dietary biogenic amines. Current Nutrition & Food Science. 2010;**6**(2):145-156

[7] Linares DM et al. Comparative analysis of the in vitro cytotoxicity of the dietary biogenic amines tyramine and histamine. Food Chemistry. 2016;**197**:658-663

[8] Kim M-K, Mah J-H, Hwang H-J. Biogenic amine formation and bacterial contribution in fish, squid and shellfish. Food Chemistry. 2009;**116**(1):87-95

[9] Hungerford JM. Scombroid poisoning: A review. Toxicon. 2010;**56**(2):231-243

[10] Bulushi IA et al. Biogenic amines in fish: Roles in intoxication, spoilage, and nitrosamine formation—A review. Critical Reviews in Food Science and Nutrition. 2009;**49**(4):369-377

[11] Ruiz-Capillas C, Jiménez-Colmenero F. Biogenic amines in meat and meat products. Critical Reviews in Food Science and Nutrition. 2005;**44**(7-8):489-599

[12] Food and Drug Administration. Fish and Fishery Products Hazards and Controls Guidance. US Department of Health and Human Services Food and Drug Administration Center for Food Safety and Applied Nutrition; 2011

[13] Directive E. Amending regulation (EC) No 2073/2005 on microbiological criteria for foodstuffs. Commission regulation (EC) No 1441/2007 of 5 December 2007. Official Journal of the European Union. 2007;**322**:12-29

[14] Rodriguez MBR et al. Bioactive amines: Aspects of quality and safety in food. Food and Nutrition Sciences. 2014;**5**(02):138

[15] Bauza T et al. Determination of biogenic amines and their precursor amino acids in wines of the Vallée du Rhône by high-performance liquid chromatography with precolumn derivatization and fluorimetric detection. Journal of Chromatography A. 1995;**707**(2):373-379

[16] Landete JM et al. Biogenic amines in wines from three Spanish regions. Journal of Agricultural and Food Chemistry. 2005;**53**(4):1119-1124

[17] Konakovsky V et al. Levels of histamine and other biogenic amines in high-quality red wines. Food Additives and Contaminants. 2011;**28**(4):408-416

[18] Halász A, Baráth Á, Holzapfel WH. The influence of starter culture selection on sauerkraut fermentation. Zeitschrift für Lebensmittel-Untersuchung und-Forschung A. 1999;**208**(5-6):434-438

[19] Erim FB. Recent analytical approaches to the analysis of biogenic amines in food samples. TrAC Trends in Analytical Chemistry. 2013;**52**:239-247

[20] Soufleros E et al. Determination of biogenic amines in Greek wines by HPLC and ultraviolet detection after dansylation and examination of factors affecting their presence and concentration. Food Chemistry. 2007;**101**(2):704-716

[21] Arce L, Ríos A, Valcárcel M. Direct determination of biogenic amines in wine by integrating continuous flow clean-up and capillary electrophoresis with indirect UV detection. Journal of Chromatography A. 1998;**803**(1-2):249-260

[22] Saccani G et al. Determination of biogenic amines in fresh and processed meat by suppressed ion chromatography-mass spectrometry using a cation-exchange column. Journal of Chromatography A. 2005;**1082**(1):43-50

[23] Pastore P et al. Determination of biogenic amines in chocolate by ion chromatographic separation and pulsed integrated amperometric detection with implemented waveform at Au disposable electrode. Journal of Chromatography A. 2005;**1098**(1-2):111-115

[24] Kvasnička F, Voldřich M. Determination of biogenic amines by capillary zone electrophoresis with conductometric detection. Journal of Chromatography A. 2006;**1103**(1):145-149

[25] Anlı RE, Bayram M. Biogenic amines in wines. Food Reviews International. 2008;**25**(1):86-102

[26] Proestos C, Loukatos P, Komaitis M. Determination of biogenic amines in wines by HPLC with precolumn dansylation and fluorimetric detection. Food Chemistry. 2008;**106**(3):1218-1224

[27] Molins-Legua C, Campins-Falcó P. Solid phase extraction of amines. Analytica Chimica Acta. 2005;**546**(2):206-220

[28] Basheer C et al. Hydrazone-based ligands for micro-solid phase extraction-high performance liquid chromatographic determination of biogenic amines in orange juice. Journal of Chromatography A. 2011;**1218**(28):4332-4339

[29] Basozabal I et al. Rational design and chromatographic evaluation of histamine imprinted polymers optimised for solid-phase extraction of wine samples. Journal of Chromatography A. 2013;**1308**:45-51

[30] Rezaee M et al. Determination of organic compounds in water using dispersive liquid-liquid microextraction. Journal of Chromatography A. 2006;**1116**(1-2):1-9

[31] Donthuan J, Yunchalard S, Srijaranai S. Vortex-assisted surfactant-enhanced-emulsification liquid-liquid microextraction of biogenic amines in fermented foods before their simultaneous analysis by high-performance liquid chromatography. Journal of Separation Science. 2014;**37**(21):3164-3173

[32] Pedersen-Bjergaard S, Rasmussen KE. Liquid-phase microextraction with porous hollow fibers, a miniaturized and highly flexible format for liquid-liquid extraction. Journal of Chromatography A. 2008;**1184**(1-2):132-142

[33] Ramos RM, Valente IM, Rodrigues JA. Analysis of biogenic amines in wines by salting-out assisted liquid-liquid extraction and high-performance liquid chromatography with fluorimetric detection. Talanta. 2014;**124**:146-151

[34] Paleologos E et al. Determination of biogenic amines as their benzoyl

derivatives after cloud point extraction with micellar liquid chromatographic separation. Journal of Chromatography A. 2003;**1010**(2):217-224

[35] Peña-Gallego A et al. High-performance liquid chromatography analysis of amines in must and wine: A review. Food Reviews International. 2012;**28**(1):71-96

[36] Pineda A et al. Preliminary evaluation of biogenic amines content in Chilean young varietal wines by HPLC. Food Control. 2012;**23**(1):251-257

[37] Henríquez-Aedo K et al. Evaluation of biogenic amines content in Chilean reserve varietal wines. Food and Chemical Toxicology. 2012;**50**(8):2742-2750

[38] Zhai H et al. Biogenic amines in commercial fish and fish products sold in southern China. Food Control. 2012;**25**(1):303-308

[39] Šimat V, Dalgaard P. Use of small diameter column particles to enhance HPLC determination of histamine and other biogenic amines in seafood. LWT-Food Science and Technology. 2011;**44**(2):399-406

[40] Huang K-J et al. Development of an ionic liquid-based ultrasonic-assisted liquid-liquid microextraction method for sensitive determination of biogenic amines: Application to the analysis of octopamine, tyramine and phenethylamine in beer samples. Journal of Chromatography B. 2011;**879**(9-10):579-584

[41] Li G et al. Simultaneous determination of biogenic amines and estrogens in foodstuff by an improved HPLC method combining with fluorescence labeling. LWT-Food Science and Technology. 2014;**55**(1):355-361

[42] Kelly MT, Blaise A, Larroque M. Rapid automated high performance liquid chromatography method for simultaneous determination of amino acids and biogenic amines in wine, fruit and honey. Journal of Chromatography A. 2010;**1217**(47):7385-7392

[43] Iijima S et al. Optimization of an online post-column derivatization system for ultra high-performance liquid chromatography (UHPLC) and its applications to analysis of biogenic amines. Analytical Sciences. 2013;**29**(5):539-545

[44] Gao PF et al. Determination of trace biogenic amines with 1, 3, 5, 7-tetramethyl-8-(N-hydroxysuccinimidyl butyric ester)-difluoroboradiaza-s-indacene derivatization using high-performance liquid chromatography and fluorescence detection. Journal of Separation Science. 2011;**34**(12):1383-1390

[45] Özdestan Ö, Üren A. A method for benzoyl chloride derivatization of biogenic amines for high performance liquid chromatography. Talanta. 2009;**78**(4-5):1321-1326

[46] Kim JY et al. Effects of storage temperature and time on the biogenic amine content and microflora in Korean turbid rice wine, Makgeolli. Food Chemistry. 2011;**128**(1):87-92

[47] Piasta AM et al. New procedure of selected biogenic amines determination in wine samples by HPLC. Analytica Chimica Acta. 2014;**834**:58-66

[48] Bach B et al. Validation of a method for the analysis of biogenic amines: Histamine instability during wine sample storage. Analytica Chimica Acta. 2012;**732**:114-119

[49] Jain A, Gupta M, Verma KK. Salting-out assisted liquid-liquid extraction for the determination of biogenic amines in fruit juices and alcoholic

beverages after derivatization with 1-naphthylisothiocyanate and high performance liquid chromatography. Journal of Chromatography A. 2015;**1422**:60-72

[50] Valente IM et al. Application of gas-diffusion microextraction for high-performance liquid chromatographic analysis of aliphatic amines in fermented beverages. Analytical Methods. 2012;**4**(8):2569-2573

[51] De Mey E et al. Dabsyl derivatisation as an alternative for dansylation in the detection of biogenic amines in fermented meat products by reversed phase high performance liquid chromatography. Food Chemistry. 2012;**130**(4):1017-1023

[52] Köse S et al. Biogenic amine contents of commercially processed traditional fish products originating from European countries and Turkey. European Food Research and Technology. 2012;**235**(4):669-683

[53] Tameem AA et al. A 4-hydroxy-N′-[(E)-(2-hydroxyphenyl) methylidene] benzohydrazide-based sorbent material for the extraction-HPLC determination of biogenic amines in food samples. Talanta. 2010;**82**(4):1385-1391

[54] Sagratini G et al. Simultaneous determination of eight underivatised biogenic amines in fish by solid phase extraction and liquid chromatography-tandem mass spectrometry. Food Chemistry. 2012;**132**(1):537-543

[55] Papageorgiou M et al. Direct solid phase microextraction combined with gas chromatography-mass spectrometry for the determination of biogenic amines in wine. Talanta. 2018;**183**:276-282

[56] Mayr CM, Schieberle P. Development of stable isotope dilution assays for the simultaneous quantitation of biogenic amines and polyamines in foods by LC-MS/MS. Journal of Agricultural and Food Chemistry. 2012;**60**(12):3026-3032

[57] Jastrzębska A et al. Application of 3, 5-bis-(trifluoromethyl) phenyl isothiocyanate for the determination of selected biogenic amines by LC-tandem mass spectrometry and 19F NMR. Food Chemistry. 2018;**239**:225-233

[58] Restuccia D et al. A new method for the determination of biogenic amines in cheese by LC with evaporative light scattering detector. Talanta. 2011;**85**(1):363-369

[59] Sun J et al. Direct separation and detection of biogenic amines by ion-pair liquid chromatography with chemiluminescent nitrogen detector. Journal of Chromatography A. 2011;**1218**(29):4689-4697

[60] Zhong J-J et al. Liquid chromatographic method for toxic biogenic amines in foods using a chaotropic salt. Journal of Chromatography A. 2015;**1406**:331-336

[61] Ordóñez JL et al. Recent trends in the determination of biogenic amines in fermented beverages—A review. Analytica Chimica Acta. 2016;**939**:10-25

[62] Nguyen DT-T et al. High throughput liquid chromatography with sub-2 μm particles at high pressure and high temperature. Journal of Chromatography A. 2007;**1167**(1):76-84

[63] Jia S et al. Simultaneous determination of 23 amino acids and 7 biogenic amines in fermented food samples by liquid chromatography/quadrupole time-of-flight mass spectrometry. Journal of Chromatography A. 2011;**1218**(51):9174-9182

[64] Jia S et al. Determination of biogenic amines in Bokbunja (*Rubus coreanus* Miq.) wines using

a novel ultra-performance liquid chromatography coupled with quadrupole-time of flight mass spectrometry. Food Chemistry. 2012;**132**(3):1185-1190

[65] Romero-González R et al. Simultaneous determination of four biogenic and three volatile amines in anchovy by ultra-high-performance liquid chromatography coupled to tandem mass spectrometry. Journal of Agricultural and Food Chemistry. 2012;**60**(21):5324-5329

[66] Lee S, Yoo M, Shin D. The identification and quantification of biogenic amines in Korean turbid rice wine, Makgeolli by HPLC with mass spectrometry detection. LWT-Food Science and Technology. 2015;**62**(1):350-356

[67] Jastrzębska A, Piasta A, Szłyk E. Application of ion chromatography for the determination of biogenic amines in food samples. Journal of Analytical Chemistry. 2015;**70**(9):1131-1138

[68] Palermo C et al. A multiresidual method based on ion-exchange chromatography with conductivity detection for the determination of biogenic amines in food and beverages. Analytical and Bioanalytical Chemistry. 2013;**405**(2-3):1015-1023

[69] Mohammed G et al. A critical overview on the chemistry, clean-up and recent advances in analysis of biogenic amines in foodstuffs. TrAC Trends in Analytical Chemistry. 2016;**78**:84-94

[70] Bilgin B, Gençcelep H. Determination of biogenic amines in fish products. Food Science and Biotechnology. 2015;**24**(5):1907-1913

[71] Romano A et al. Determination of biogenic amines in wine by thin-layer chromatography/densitometry. Food Chemistry. 2012;**135**(3):1392-1396

[72] Önal A. A review: Current analytical methods for the determination of biogenic amines in foods. Food Chemistry. 2007;**103**(4):1475-1486

[73] Almeida C, Fernandes J, Cunha S. A novel dispersive liquid-liquid microextraction (DLLME) gas chromatography-mass spectrometry (GC-MS) method for the determination of eighteen biogenic amines in beer. Food Control. 2012;**25**(1):380-388

[74] Cunha S, Faria M, Fernandes J. Gas chromatography-mass spectrometry assessment of amines in port wine and grape juice after fast chloroformate extraction/derivatization. Journal of Agricultural and Food Chemistry. 2011;**59**(16):8742-8753

[75] Płotka-Wasylka J, Simeonov V, Namieśnik J. An in situ derivatization-dispersive liquid-liquid microextraction combined with gas-chromatography-mass spectrometry for determining biogenic amines in home-made fermented alcoholic drinks. Journal of Chromatography A. 2016;**1453**:10-18

[76] Huang J et al. Environmentally friendly solid-phase microextraction coupled with gas chromatography and mass spectrometry for the determination of biogenic amines in fish samples. Journal of Separation Science. 2016;**39**(22):4384-4390

[77] Guo YY et al. Biogenic amines in wine: A review. International Journal of Food Science and Technology. 2015;**50**(7):1523-1532

[78] Uzaşçı S, Başkan S, Erim FB. Biogenic amines in wines and pomegranate molasses—A non-ionic micellar electrokinetic chromatography assay with laser-induced fluorescence detection. Food Analytical Methods. 2012;**5**(1):104-108

[79] Zhang N et al. Sensitive determination of biogenic amines

by capillary electrophoresis with a new fluorogenic reagent 3-(4-fluorobenzoyl)-2-quinolinecarboxaldehyde. Talanta. 2008;**76**(4):791-797

[80] Daniel D et al. Determination of biogenic amines in beer and wine by capillary electrophoresis-tandem mass spectrometry. Journal of Chromatography A. 2015;**1416**:121-128

[81] Ginterová P et al. Determination of selected biogenic amines in red wines by automated on-line combination of capillary isotachophoresis-capillary zone electrophoresis. Journal of Chromatography B. 2012;**904**:135-139

[82] Jastrzębska A, Piasta A, Szłyk E. Simultaneous determination of selected biogenic amines in alcoholic beverage samples by isotachophoretic and chromatographic methods. Food Additives & Contaminants: Part A. 2014;**31**(1):83-92

[83] Dossi N et al. A modified electrode for the electrochemical detection of biogenic amines and their amino acid precursors separated by microchip capillary electrophoresis. Electrophoresis. 2011;**32**(8):906-912

[84] Grieshaber D et al. Electrochemical biosensors—Sensor principles and architectures. Sensors. 2008;**8**(3):1400-1458

[85] Apilux A et al. Paper-based acetylcholinesterase inhibition assay combining a wet system for organophosphate and carbamate pesticides detection. EXCLI Journal. 2015;**14**:307-319

[86] Ludwig SKJ et al. Calling biomarkers in milk using a protein microarray on your smartphone. PLoS One. 2015;**10**(8):e0134360

[87] Sheng W et al. Development of an enzyme-linked immunosorbent assay for the detection of tyramine as an index of freshness in meat and seafood. Journal of Agricultural and Food Chemistry. 2016;**64**(46):8944-8949

[88] Chen P et al. A novel chemiluminescence enhanced method for determination of putrescine in shrimp based on the luminol-$[Ag(HIO_6)_2]^{5-}$ reaction. Analytical Methods. 2016;**8**(5):1151-1156

[89] Gajjala RKR, Palathedath SK. Cu@ Pd core-shell nanostructures for highly sensitive and selective amperometric analysis of histamine. Biosensors and Bioelectronics. 2018;**102**:242-246

[90] Costa DJE et al. Determination of tryptamine in foods using square wave adsorptive stripping voltammetry. Talanta. 2016;**154**:134-140

[91] Kaur N et al. Fe(III) conjugated fluorescent organic nanoparticles for ratiometric detection of tyramine in aqueous medium: A novel method to determine food quality. Food Chemistry. 2018;**245**:1257-1261

[92] Yurova NS et al. Functional electrospun nanofibers for multimodal sensitive detection of biogenic amines in food via a simple dipstick assay. Analytical and Bioanalytical Chemistry. 2018;**410**(3):1111-1121

[93] Kochana J et al. Mesoporous carbon-containing voltammetric biosensor for determination of tyramine in food products. Analytical and Bioanalytical Chemistry. 2016;**408**(19):5199-5210

[94] Dong X-X et al. Portable amperometric immunosensor for histamine detection using Prussian blue-chitosan-gold nanoparticle nanocomposite films. Biosensors and Bioelectronics. 2017;**98**:305-309

[95] Ramon-Marquez T et al. Novel optical sensing film based on a

functional nonwoven nanofiber mat for an easy, fast and highly selective and sensitive detection of tryptamine in beer. Biosensors and Bioelectronics. 2016;**79**:600-607

[96] Li Y et al. Tyramine detection using PEDOT:PSS/AuNPs/1-methyl-4-mercaptopyridine modified screen-printed carbon electrode with molecularly imprinted polymer solid phase extraction. Biosensors and Bioelectronics. 2017;**87**:142-149

[97] Ansorena D et al. Analysis of biogenic amines in northern and southern European sausages and role of flora in amine production. Meat Science. 2002;**61**(2):141-147

[98] Visciano P et al. Biogenic amines in raw and processed seafood. Frontiers in Microbiology. 2012;**3**:188

Section 3

Biogenic Amines in Wines

Chapter 3

Biogenic Amines: A Claim for Wines

Maria Martuscelli and Dino Mastrocola

Abstract

Many possible factors influence the accumulation of biogenic amines in wines, correlated both to agronomical practices in the vineyard and during the winemaking process. In the literature, it is reported that the quantities of biogenic amines found in many wines are not alarming, especially with regard to those of toxicological interest (histamine and tyramine). For subjects in specific physiological conditions (histamine intolerance, taking class of drugs that inhibit monoamine oxidase enzymes), the risk of creating toxic reactions is related to the composition of the whole meal, not only the consumption of wine. It would be desirable to establish a regulatory system, as already existing for sulphites, allowing to read a label with the claim specifying their absence (e.g., histamine free) in order to enhance the quality of wines that would be a priori forbidden.

Keywords: biogenic amines, wine, winemaking process, quality, safety

1. Introduction

The level of biogenic amines (BA) in wines is an important quality and safety parameter, in addition to the composition in polyphenols and polysaccharides [1, 2].

The presence of biogenic amines in wine is quite frequent and inevitable; they are naturally present in grapes but also deriving from the decarboxylation of amino acids by enzymes of microbial origin, or even produced by transamination of aldehydes by amino acid transaminase [3–6].

In most of the studies carried out, more amines have been observed in red wine than either white or rosé ones [7]. There is a high correlation among chemical and microbiological characteristics of grape variety, winemaking process conditions, and amount of the biogenic amines in wines [5, 8].

Histamine, tyramine, and cadaverine biogenic amines are most representative of the wine; other amines such as putrescine, ethylamine, 2-phenylethylamine, spermine, and spermidine are also present in the grape must [9].

Some factors of agronomic practice as well as of the winemaking process can cause discrete levels of biogenic amines in the wine: fertilization of the soil (nitrogen level), poor state of health of the grapes and presence of molds, nonregular lowering of the pH of the must and development of non-*Saccharomyces* yeasts, and activity of lactic acid bacteria responsible for secondary fermentation (malolactic fermentation, MLF) [8, 10].

Many technological, biological, and environmental factors can affect the occurrence of biogenic amines in wine such as skin maceration or post-fermentative maceration or contact with the lees [11–13].

The occurrence of biogenic amines in wines is also a consequence of the various treatments applied in winemaking process, some of them favoring the amines synthesis (wine and must treatments with yeast mannoproteins or proteolytic enzymes) [14, 15], and others, such as the use of clarification substances and oenological adjuvants (bentonite and polyvinylpolypyrrolidone), are instead able to absorb biogenic amines and then lower their levels in the final products [4].

In addition, mold infections of grapes display significant impacts on the initial content of biogenic amines in grape must, and their level is not efficiently reduced from fining agents currently available for the wine industry [16].

2. Why know the biogenic amines level in wines?

The knowledge of the biogenic amine levels in wines is important to both consumers and producers. Some authors proposed biogenic amine content as markers for authentication of origin in addition to other chemical compounds as polyphenols [3, 17].

Biogenic amines are present as salts but, at the pH prevailing of mouth, they are partly in free form, becoming reactive with other compounds responsible for the aroma of the wine so to impart aromatic notes own or even indirectly; they can be responsible of sensory changes and occult (e.g., loss of varietal character) and overt (musty smell of tuna) [18, 19].

But what makes it very useful to know the level of amines in wines is related to the fact that they are generally recognized among the most important causes for the intolerance to wine originating intoxication symptoms as nausea, cardiac palpitations, headaches, flushing, and increase or decrease blood pressure, as also reported by European Food Safety Authority [20, 21]. High sensitivity toward biogenic amines ingested with the diet depends on insufficient amino oxidase activity caused by drugs, genetic predisposition (histamine intolerance), gastrointestinal disease, inhibition by alcohol, acetaldehyde, and other amines (e.g., putrescine and cadaverine) [7].

Monitoring of the amine levels in wines can be an important marketing advantage. A large data set (from various production sites, vintage years, etc.) is necessary to establish BA profiles as wine fingerprint and provide scientific basis for safety and quality control in winemaking process.

2.1 Occurrence of biogenic amines in the red, white, and rosè wines

Different processes are applied for the production of red, white, or rosé wines, in which the chemical and physico-chemical characteristics result different and differently correlated with their biogenic amine content [22].

It is well known that SO_2 is added in higher concentration in white wine to ensure color stabilization, particularly in the post-malolactic fermentation and before bottling, therefore the total SO_2 value can be rather variable, but it must be in any case below the legal limits. In red wines, higher pH values than white or rosè ones are a consequence of malolactic fermentation in which lactic acid bacteria degrade malic acid to lactic acid and CO_2 (biological deacidification). A parameter of wine very useful for predicting the amino biogenic activity can be pH, especially considering that one of the explanations is that microorganisms use the biogenic amines synthesis as a metabolic strategy in response to environmental acidity or as a source of alternative energy [23].

As regard alcohol degree that is considered one of the parameters affecting the activities of amino acid decarboxylating bacteria involved in MLF, it is lower in white and rosé wines compared to red ones [24]. Generally, red wines with maximum alcohol content (about 14.5%) belong to the Reserve "Controlled Designation of Origin" (CDO) wines.

According to Bauer and Dicks [25], the malic acid content indicates a complete MLF (occurring in all red wines) or an incomplete/absent MLF (in white and rosè wines). Many studies were carried out on the role of MLF on the content of biogenic amines in wines and on the factors that can affect their production [11, 24, 26–28]. The variability of the biogenic amines distribution in red, rosé, and white wines is remarkable and could be attributed to the numerous variables affecting biogenic amine formation by bacteria during vinification and wine storage.

High amounts of diamine putrescine as well as the biologically active amines histamine and tyramine are determined in red wines, higher than those found in rosé and white wine [2, 29–32]. Generally, red wine vinification is carried out in the presence of grape skin and pulp, and putrescine could be released from these into the must [33].

The presence of putrescine has been associated with the secondary fermentation (MLF) and in particular with the activity of *O. oeni* strains although it has been demonstrated that *L. hilgardii* and *L. plantarum* are able to produce putrescine during alcoholic fermentation, using substrates from autolysis of yeast wine strains [34–36].

During vinification, putrescine can also originate from the microbial decarboxylation of ornithine or from arginine via agmatine [37]. Furthermore, some study demonstrated that amines are formed during alcoholic fermentation, even from the very beginning, involving yeasts and molds [38].

Moreover, high histamine and cadaverine levels in wines in which oak barrels were employed for secondary fermentation MLF [22].

In wines, polyamines might originate from grapes and/or from yeast lysis [33]; therefore the low amount of spermidine in rosè wines could be explained taking into account that yeasts are enable to liberate polyamines in significant amount. A significantly decreasing of endogenous spermidine can be observed during winemaking, due to the growth of lactic acid bacteria, to potential consumption by alcoholic fermentative yeasts or to spontaneous chemical degradation [30, 31].

Moreover, higher levels of biogenic amines in the CDO reserve wines could be explained by the influence of the aging treatments [2, 12]. Cells of some lactic bacteria surviving aging are still able to decarboxylate amino acids and, due to the increase in pH value, multiply giving rise to the biogenic amine accumulation [26]. A relative high amount of biogenic amines in a Sherry-like wine as Vernaccia di Oristano could be explained with the long aging process, respect to those of newly made white wines [2, 39].

A separate case is the liqueurs, whose process is completely different from that of winemaking [40]; however, being counted among the alcoholic beverages consumed in a meal or in any case during the day, it could be useful to know the case also with respect to these products. For example, in [41], it is reported that high levels of BA in coffee, honey, and fruits liqueurs are significantly higher than those occurred in milk and herb liqueurs. The variability observed between samples was influenced by the type of components as well as by the different modes of production (homemade or industrial). Indeed, homemade sample had significantly higher amounts of BA than industrial samples. To date, no studies have been reported on the content of biogenic amines in homemade wines. Since this practice is still widely used in many regions of Italy, it would be desirable to collect data on the matter.

Significant seems to be the effect of the winery, regardless of the geographic area where it is located, in according to the results of other authors [15]. A predominance of decarboxylating positive microbiota can occur in some wineries, even if the fermentation is carried out by commercial starter strains [9].

However, great efforts are made in recent years by winemakers to reduce the content of amines and improve the quality and safety of wines: a study carried out on about 700 samples of Spanish wine, demonstrated that the average values of the histamine content for red, rosé, and white wines decreased, from 2010 to 2015, and in any case, none of the samples exceeded 10 mg L^{-1} of histamine [1].

2.2 Exposure to biogenic amine human intake via wine consumption: a case study in China

The first study at the national level to assess and link the BA contents to potential health risks for the largest group of wine consumers was recently carried out in China by Ke et al. [42]. Average annual per capita wine consumption in China is 0.75–1.2 L, whereas it is 54 and 49 L in France and Italy, respectively [43]; so, two different exposure scenarios can be designed: chronic and acute exposures. A general physiologically based pharmacokinetic (PBPK) model was used in order to estimate the tissue concentrations of biogenic amines in humans after intake via wine consumption [42]. In this study, the chronic exposure scenario was equivalent of consumption per person of 150 mL (one glass) of wine per day through a whole year (365 days). This corresponded to 54.8 L of wine in a year. In the acute exposure scenario, it was assumed that the person intakes 750 mL of wine (one bottle) in a single oral dose; the BA fate in the human body was simulated for 30 days. The dietary intake amount of biogenic amines in both exposure scenarios was calculated by using the median concentrations of BAs (putrescine, cadaverine, histamine, tyramine, phenylethylamine, spermidine, and spermine) detected in 456 wines sold in China (>90% of wine brands in a three-year span). The simulation considers the simultaneous uptake of a group of BAs together with alcohol, a situation that may inhibit BA metabolism. The PBPK model simulations were carried out by assuming no metabolism/transformation of BAs, simulating the pharmacokinetic fate of BAs for the most sensitive population in a critical exposure scenario. Results of this study demonstrated that in a chronic exposure scenario, the steady-state mass fractions of phenylethylamine and tyramine present in the adipose tissues are 77 and 65%, respectively, remarkably higher than those of other biogenic amines. The highest steady-state mass fractions of the other biogenic amines were found in the muscle tissues, with a 40% of the total body mass, while the mass fractions of blood, adipose, and liver tissues are 7.4, 21 and 2.6%, respectively. In the acute exposure scenario, the entire dose of amine was lost from the body at the end of the 30-day simulation; in particular, it took more than a week for the complete loss of phenylethylamine, and about one week was required for tyramine, whereas it was about one day for all the simulated biogenic amines. For all the simulated BAs, urinary loss accounted for >99% of the loss processes, whereas fecal and air loss were negligible (<1%).

2.3 Defining a biogenic amine safety threshold for wines

Actually, the European Union (EU) has not established regulations for the wine industry, but it has only suggested the "safety threshold values" [6].

Wines imported into EU are often accompanied by a certificate of analysis regarding biogenic amine content, even if there is no EU regulation about their limit in wines; however, Germany, Belgium, and France have differently recommended not to exceed histamine levels of 2, 6, 8 mg L^{-1}, respectively, while Switzerland has removed its official maximum histamine limit (previously set at 10 mg L^{-1}) in imported wines [44]. The biogenic amine content in wines must be interpreted in the light of the toxicology studies on consumers [21], considering that biogenic amines have different safety threshold values depending on the physiological conditions of consumers. In general, the amines taken with the diet are metabolized by the activity of enzymes located in the intestine, the mono- and di-amino oxidases (MAO and DAO, respectively). The activity of intestinal MAO and DAO can be inhibited by alcohol and by some antidepressant and hypotensive drugs, the so-called MAO inhibitors (MAOIs). Diamine (putrescine and cadaverine) and polyamine (spermine and spermidine) are histamine, tyramine, tryptamine, and phenylethylamine toxicity enhancers, as they compete in the detoxification metabolism carried out by MAO and DAO. For subjects having histamine intolerance (genetic deficiency in diamine oxidase), patients taking MAOI drugs or patients under new generation MAOI treatment, so called RIMA (reversible inhibitors of MAO-A drugs), the risk of creating toxic reactions is related to the composition of the whole meal, not only the consumption of wine.

Table 1 shows the amounts of biogenic amines found in 100 mL of commercial wines classified as "Controlled Designation of Origin" (CDO) and "Typical Geographical Indication" (TGI) (Commission Regulation EC No 628/2008) of different regions of South Italy. Results shown in **Table 1** demonstrate that the levels of biogenic amines can be not alarming in the commercial wines, especially with regard to histamine and tyramine, the most important cause of intolerance to wine [20].

Commercial wines	grape variety/ies	Putrescine	Cadaverine	Histamine	Tyramine
Montepulciano d'Abruzzo CDO [2]	100% Montepulciano	0,789±0,602	0,012±0,030	0,311±0,275	0,551±0,493
Primitivo del Salento TGI [2]	100% Primitivo	0,870±0,128	n.d.	0,055±0,021	0,181±0,255
Negramaro TGI [2]	100% Negramaro	0,689±0,130	n.d.	0,040±0,018	0,362±0,22
Aglianico del Vulture CDO [3]	100% Aglianico	0,324±0,118	0,174±0,104	0,195±0,113	0,261±0,141
Aglianico Irpinia CDO [3]	100% Aglianico	0,223±0,070	0,321±0,124	0,022±0,009	0,114±0,004
Gaglioppo [3]	100% Gaglioppo	0,599±0,331	0,194±0,051	0,301±0,106	0,122±0,016
Montepulciano d'Abruzzo Cerasuolo CDO [2]	100% Montepulciano	0,339±0,246	0,090±0,014	0,065±0,128	0,097±0,274
Faro CDO [32]	60% Nerello Mascalese, 30% Nerello Cappuccio, 10% Corinto and Nero D'Avola	0,233±0,005	0,004±0,000	0,123±0,001	0,045±0,001
Nero D'Avola TGI [32]	100% Nero D'Avola	0,033±0,000	0,002±0,000	0,152±0,002	0,264±0,006
Chardonnay Terre di Chieti TGI [2]	100% Chardonnay	0,150±0,071	n.d.	n.d.	n.d.
Trebbiano d'Abruzzo TGI [2]	100% Trebbiano	0,267±0,334	0,100±0,000	0,025±0,087	0,008±0,029
Mamertino TGI [32]	45% Catarratto, 20% Inzolia, 15% Grillo, 20% other	0,122±0,001	n.d.	0,007±0,000	0,014±0,000
Trebbiano TGI [32]	100% Trebbiano	0,054±0,001	0,004±0,000	0,006±0,000	0,005±0,000
Inzolia TGI [32]	100% Inzolia	0,126±0,001	n.d.	0,004±0,000	n.d.
Vernaccia di Oristano CDO [39]	100% Vernaccia	1,33±0,88	0,060±0,03	1,22±0,49	0,31±0,18

Table 1.
Biogenic amine contents (mg/100 mL) in commercial wines of South Italy.

The statistical distribution of histamine in 684 samples of Spanish wines has been modeled using the β-content tolerance intervals [1]. Besides, copulas to obtain the joint multivariate confidence region between histamine and tyramine have been built for the first time in the oenological field. Authors demonstrated that it is necessary for the nonparametric approach to correctly analyze the distribution of BA content in wines, because the data do not follow a normal distribution nor their relations are linear; in the univariate approach, the β-content tolerance interval permits to obtain the proportion of wines that meet a "possible" limit.

3. Conclusions

In order to evaluate the real risks of biogenic amine intake associated with the consumption of wines, it must be considered that a chronic exposure scenario can be equivalent of consumption per person of 150 mL (one glass) of wine per day; moreover, in the Mediterranean diet, the consumption of wine with meals can be about 100–200 mL (for woman and man consumer, respectively).

It should be consider that the consumer needs a positive message about the product, so that it can be encouraged to purchase. On the other hand, it is necessary that the consumer is correctly informed of the content of potentially toxic compounds, also based on its physiological and pathological status. In general, wine is forbidden a priori to sensitive consumers, regardless of its real quality and safety. Therefore, if the wine label shows information about the composition in biogenic amines, both qualitative and quantitative, it would make the consumer aware of the real risk and then a double goal would be achieved: avoiding unpleasant or even harmful effects on his health, allowing the consumption of wine low or completely free of amines with a toxic effect. It would therefore be desirable that, after extensive collection of toxicological data and following an extensive media information campaign, establish for the wines a regulatory system, as already existing for sulphites, which allows the communication of the values of amines present, or, again better, a statement that can enhance the wines in which these compounds are absent, for example, a claim *"histamine free."*

Conflict of interest

The author, the undersigned corresponding author, declares that we have no commercial associations that might pose a conflict of interest in connection with the submitted chapter.

Biogenic Amines: A Claim for Wines
DOI: http://dx.doi.org/10.5772/intechopen.80362

Author details

Maria Martuscelli* and Dino Mastrocola
Faculty of Bioscience and Technology for Food, Agriculture and Environment,
University of Teramo, Teramo, Italy

*Address all correspondence to: mmartuscelli@unite.it

IntechOpen

References

[1] Melendez ME, Sarabia LA, Ortiz MC. Distribution free methods to model the content of biogenic amines in Spanish wines. Chemometrics and Intelligent Laboratory Systems. 2016;**155**:191-199

[2] Martuscelli M, Arfelli G, Manetta AC, Suzzi G. Biogenic amines content as a measure of the quality of wines of Abruzzo (Italy). Food Chemistry. 2013;**140**:590-597

[3] Galgano F, Caruso M, Perretti G, Favati F. Authentication of Italian red wines on the basis of the polyphenols and biogenic amines. European Food Research and Technology. 2011;**232**:889-897

[4] Pozo-Bayon MA, Monagas M, Bartolomè B, Moreno-Arribas V. Wine futures related to safety and consumer health: An integrated perspective. Critical Reviews in Food Science and Nutrition. 2012;**52**(1):31-54

[5] Del Prete V, Costantini A, Cecchini F, Morassut M, Garcia-Moruno E. Occurrence of biogenic amines in wine: The role of grapes. Food Chemistry. 2009;**112**:474-481

[6] Karovičová J, Kohajdová Z. Biogenic amines in food. Chemical Papers. 2005;**59**(1):70-79

[7] Guo YY, Yang YP, Peng Q, Han Y. Biogenic amines in wine: A review. International Journal of Food Science and Technology. 2015;**50**(7):1523-1532

[8] Ancín-Azpilicueta C, González-Marco A, Jiménez-Moreno N. Current knowledge about the presence of amines in wine. Critical Reviews in Food Science and Nutrition. 2008;**48**:257-275

[9] Costantini A, Vaudano E, Del Prete V, Danei M, Garcia-Moruno E. Biogenic amine production by contaminating bacteria found in starter preparations used in winemaking. Journal of Agricultural and Food Chemistry. 2009;**57**(22):10664-10669

[10] OIV Guides. OIV Code of Good Vitivinicultural Practices in Order to Minimise the Presence of Biogenic Amines in Vine-Based Products. CST 369-2011; 2011

[11] Alcaide-Hidalgo JM, Moreno-Arribas MV, Martín-Álvarez PJ, Polo MC. Influence of malolactic fermentation, postfermentative treatments and ageing with lees on nitrose compounds of red wines. Food Chemistry. 2007;**103**:572-581

[12] Hernández-Orte P, Lapeña AC, Peña-Gallego A, Astrain J, Baron C, Pardo I, et al. Biogenic amine determination in wine fermented in oak barrels. Factors affecting formation. Food Research International. 2008;**41**:697-706

[13] Pérez-Serradilla JA, Luque de Castro MD. Role of lees in wine production: A review. Food Chemistry. 2008;**111**:447-456

[14] García-Marino M, Trigueros Á, Escribano-Bailón T. Influence of oenological practices on the formation of biogenic amines in quality red wines. Journal of Food Composition and Analysis. 2010;**23**:455-462

[15] Martín-Álvarez PJ, Marcobal A, Polo C, Moreno-Arribas MV. Influence of technological practices on biogenic amine contents in red wine. European Food Research and Technology. 2006;**222**:420-424

[16] Grossmann M, Smit I, Loehnertz O, Ansorge A. Biogenic amines and grapes: Effect of microbes and fining agents. In: Proceeding of International Symposium of Microbiology and Food

Safety of Wine; 20-21 November 2007; Vilafranca, Spain

[17] Herbert P, Cabrita MJ, Ratoa N, Laureano O, Alves A. Free amino acids and biogenic amines in wines and musts from the Alentejo region. Evolution of amines during alcoholic fermentation and relationship with variety, sub-region and vintage. Journal of Food Engineering. 2005;**66**:315-322

[18] Cappello MS, Zapparoli G, Logrieco A, Bartowsky EJ. Linking wine lactic acid bacteria diversity with wine aroma and flavour. Review article. International Journal of Food Microbiology. 2017;**243**:16-27

[19] Smit AY, du Toit WJ, du Toit M. Biogenic amines in wine: Understanding the headache. South African Journal of Enology and Viticulture. 2008;**29**(2):109-127

[20] Konakovsky V, Focke M, Hoffmann-Sommergruber K, Schmid R, Scheiner O, et al. Levels of histamine and other biogenic amines in high-quality red wines. Food additives & contaminants. Food Additives & Contaminants: Part A: Chemistry, Analysis, Control, Exposure & Risk Assessment. 2011;**28**(4):408-416

[21] Scientific Opinion on risk based control of biogenic amine formation in fermented foods. EFSA Journal. 2011;**9**(10):2393

[22] Płotka-Wasylka J, Simeonov V, Morrison C, Namieśnik J. Impact of selected parameters of the fermentation process of wine and wine itself on the biogenic amines content: Evaluation by application of chemometric tools. Microchemical Journal. 2018;**142**:187-194

[23] Cotter PD, Hill C. Surviving the acid test: Responses of Gram-positive bacteria to low pH. Microbiology and Molecular Biology Reviews. 2003;**67**:429-453

[24] Coton E, Torlois S, Bertrand A, Lonvaud-Funel A. Amines biogènes et bactéries lactiques du vin. Bulletin O.I.V. 1999;**72**:23-35

[25] Bauer R, Dicks LMT. Control of malolactic fermentation in wine: A review. South African Journal of Enology and Viticulture. 2004;**25**(2):74-88

[26] Lonvaud-Funel A. Biogenic amines in wine: Role of lactic acid bacteria. FEMS Microbiology Letters. 2001;**199**:9-13

[27] Pramateftaki PV, Metafa M, Kallthraka S, Lanaridis P. Evolution of malolactic bacteria and biogenic amines during spontaneous malolactic fermentations in a Greek winery. Letters in Applied Microbiology. 2006;**43**(2):155-160

[28] Capozzi V, Russo P, Beneduce L, Weidmann S, Grieco F, Guzzo J, Spano G. Technological properties of *Oenococcus oeni* strains isolated from tipical southern italian wines. Letters in Applied Microbiology. 2010;**50**:327-334

[29] Soleas GJ, Dam J, Carey M, Goldberg DM. Toward the fingerprinting of wines: Cultivar-related patterns of polyphenolic constituents in Ontario wines. Journal of Agricultural and Food Chemistry. 1999;**45**:3871-3880

[30] Landete JM, Ferrer S, Polo L, Pardo I. Biogenic amines in wines from three Spanish regions. Journal of Agricultural and Food Chemistry. 2005;**53**:1119-1124

[31] Bover-Cid S, Iquierdo-Pulido M, Mariné-Font A, Vidal-Carou MC. Biogenic mono- di- and polyamine contents in Spanish wines and influence of a limited irrigation. Food Chemistry. 2006;**96**:43-47

[32] La Torre GL, Rando R, Saitta M, Alfa M, Maisano R, Dugo

G. Determination of biogenic amine and heavy metal contents in Sicilian wine samples. Italian Journal of Food Science. 2010;**22**:29-40

[33] Broquedis M, Dumery B, Boucard J. Mise en evidence de polyamines (putrescine, cadaverine, nor-spermidine, spermidine, spermin) dans les feuilles et le grappes de Vitis Vinifera. Connaissance de la vigne et du vin. 1989;**23**:1-6

[34] Gardini F, Zaccarelli A, Belletti N, Faustini F, Cavazza A, Martuscelli M, Mastrocola D, Suzzi G. Factors influencing biogenic amine production by a strain of *Oenococcus oeni* in a model system. Food Control. 2005;**16**(7):609-616

[35] Nannelli F, Claisse O, Gindreau E, de Revel G, Lonvaud-Funel A, Lucas PM. Determination of lactic acid bacteria producing biogenic amines in wine by quantitative PCR methods. Letters in Applied Microbiology. 2008;**47**:594-599

[36] Arena ME, Manca de Nadra MC. Biogenic amine production by Lactobacillus. Journal of Applied Microbiology. 2001;**90**:158-162

[37] Guerrini S, Mangani S, Granchi L, Vincenzini M. Biogenic amine production by *Oenococcus oeni*. Current Microbiology. 2002;**44**:374-378

[38] Grossmann M, Smit I, Loehnertz O, Ansorge A. Biogenic amines and grapes: Effect of microbes and fining agents. In: Proceeding of "International Symposium of Microbiology and food Safety of wine"; 20-21 November 2007; Vilafranca. Spain; 2007

[39] Tuberoso GIC, Serreli G, Montoro P, D'Urso G, Congiu F, Kowalczyk A. Biogenic amines and other polar compounds in long aged oxidized Vernaccia di Oristano white wines. Food Research International. 2018;**111**:97-103

[40] Regulation (EC) No 110/2008 of the European Parliament and the Council of 15 January 2008 on the Definition, Description, Presentation, Labelling and Protection of Geographical Indications of Spirit Drinks and repealing Council Regulation (EEC) No 1576/89

[41] Cunha SC, Lopes R, Fernandes JO. Biogenic amines in liqueurs: Influence of processing and composition. Journal of Food Composition and Analysis. 2017;**56**:147-155

[42] Ke R, Wei Z, Bogdal C, Göktaş RK, Xiao R. Profiling wines in China for the biogenic amines: A nationwide survey and pharmacokinetic fate modelling. Food Chemistry. 2018;**250**:268-275

[43] Qing P, Xi AQ, Hu WY. Self-consumption, gifting, and Chinese wine consumers. Canadian Journal of Agricultural Economics/Revue canadienne d'agroeconomie. 2015;**63**(4):601-620

[44] Lehtonen P. Determination of amines and amino acids in wine: A review. American Journal of Enological Viticolture. 1996;**47**(2):127-133

Biogenic Amines and Food Safety

Chapter 4

Histamine and Other Biogenic Amines in Food. From Scombroid Poisoning to Histamine Intolerance

Oriol Comas-Basté, Maria Luz Latorre-Moratalla,

Sònia Sánchez-Pérez, Maria Teresa Veciana-Nogués and

Maria del Carmen Vidal-Carou

Abstract

Histamine is a biogenic amine involved in important physiological activities in the organism, but its ingestion through food is associated with the onset of health disorders. Histamine intoxication, previously known as scombroid fish poisoning, is caused by the intake of foods with high levels of histamine. According to official European Union reports, more than 90% of the outbreaks registered in the last years were caused by the consumption of fish and seafood products. Histamine intolerance, on the other hand, arises when histamine degradation is impaired, mainly by a lower diamine oxidase (DAO) activity. Some of the uncertainties classically associated with histamine intoxication may be explained by this enzymatic deficit in a sensitive population. This chapter reviews the adverse effects of histamine from food within a risk analysis framework, focusing specifically on the components of risk assessment and management.

Keywords: histamine, biogenic amines, diamine oxidase (DAO), histamine intoxication, histamine intolerance, decarboxylase activity, risk assessment, risk management

1. Introduction

High amounts of biogenic amines in food are considered undesirable micro-components from the safety perspective, due to their potentially negative effects on consumer health. According to the risk assessment of biogenic amines carried out by the European Food Safety Authority (EFSA) [1], the amine content currently found in foods could be responsible of the triggering of health disorders. Histamine is the biogenic amine most commonly associated with the onset of health complaints. In fact, the triggering of symptoms derived by an excessive consumption of this amine was described for the first time over 60 years ago. It was firstly called scombroid fish poisoning, because the symptoms appeared mainly after the consumption of fish from the *Scombridae* and *Scomberesocidae* families, which have naturally high histidine contents. However, the World Health Organisation (WHO) recommendeds using the term histamine poisoning or histamine intoxication, since other foods can also be involved [2]. Histamine intoxication occurs in the form of an outbreak, affecting those who have consumed a particular histamine-rich food. A few years ago, another histamine-related disorder began to be described, known as histamine intolerance, which arises from the failure of the diamine oxidase (DAO) enzyme to metabolise histamine in the intestines. This enzymatic deficit in a sensitive population may explain some of the uncertainties classically associated with histamine intoxication. In this chapter we review the available information on dietary histamine and its adverse effects, using a risk analysis approach, focusing specifically on the components of risk assessment and management.

2. Risk assessment

2.1 Hazard identification

Histamine (2-[4-imidazolyl] ethylamine) is the causative agent of both histamine intoxication and histamine intolerance. Based on its chemical structure and number of amine groups, histamine is classified as a heterocyclic diamine. Important physiological activities of histamine in the human organism include synaptic transmission, blood pressure control, allergic response and cellular growth control [1]. Histamine is also found in foods, mainly fish, fish products and fermented foodstuffs [1, 3]. The major pathway for the formation of histamine in foods is the decarboxylation of its precursor amino acid, histidine, by the action of the bacterial enzyme L-histidine decarboxylase (**Figure 1**). This enzyme requires pyridoxal-5'-phosphate (vitamin B6) as a cofactor, an exception being the pyruvoyl-dependent histidine decarboxylase of Gram-positive bacteria [3–5]. Other biogenic amines commonly found in foods are tyramine, putrescine and cadaverine and to a lesser extent β-phenylethylamine and tryptamine [3, 4, 6]. These amines are all synthesised by the microbial decarboxylation of their corresponding precursor amino acids [7]. Therefore, the accumulation of histamine and other biogenic amines in foods requires the concurrence of several factors: microorganisms with decarboxylase enzymes, the availability of precursor amino acids and favourable environmental conditions for the growth or activity of aminogenic microorganisms [6, 8].

Although the ability to decarboxylate certain amino acids is a strain-dependent property, several genera of both Gram-positive and Gram-negative bacteria associated with food spoilage and/or with technological applications can produce histamine (**Table 1**) [1, 5, 9]. *Enterobacteriaceae*, including mesophilic and psychrotolerant species of *Morganella*, *Enterobacter*, *Hafnia*, *Proteus* and *Photobacterium*, are the most prolific histamine-producing bacteria in fish [4, 7, 9, 10]. In the case of fermented foods, aside from certain strains of enterobacteria, many lactic acid bacteria are also reported as histamine-producing microorganisms [1, 9].

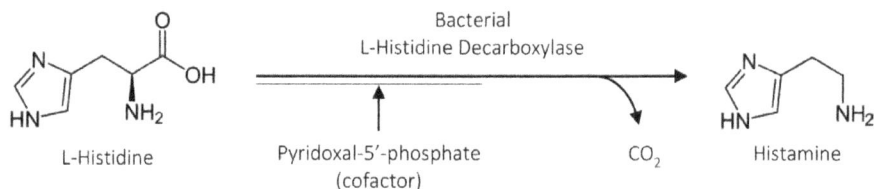

Figure 1.
Histamine formation by histidine decarboxylation.

Food	Microorganisms
Fish	*Morganella morganii, Morganella psychrotolerans, Klebsiella pneumonia, Hafnia alvei, Proteus vulgaris, Proteus mirabilis, Enterobacter cloacae, Enterobacter aerogenes, Serratia fonticola, Serratia liquefaciens, Citobacter freundii, Clostridium sp, Pseudomonas fluorescens, Pseudomonas putida, Aeromonas spp., Pleisomonas shigelloides, Photobacterium phosphoreum, Photobacterium psychrotolerans*
Fermented food (cheese, dry-fermented meat sausage, wine)	*Lactobacillus buchneri, Oenococcus oeni, Lactobacillus hilgardii, Pediococcus parvulus, Pediococcus damnosus, Lactobacillus bavaricus, Lactobacillus brevis, Lactobacillus curvatus, Lactobacillus parabuchneri, Lactobacillus rossiae, Lactobacillus saerimneri 30a, E. aerogenes, E. cloacae, Escherichia coli, H. alvei, Klebsiella oxycata, M. morganii, S. liquefaciens, Tetragenococcus spp., Leuconostoc spp.*

Table 1.
Histamine-producing microorganisms [1, 10].

The function of decarboxylase enzymes in bacterial metabolism is not fully understood, although it has been described as one of the primary emergency systems involved in the acid stress response [6]. Decarboxylases work in cooperation with a membrane antiporter protein, thus enabling amino acids to be transported into the cell and biogenic amines to be excreted out of the cell. Since decarboxylation consumes a proton, biogenic amine formation contributes to the regulation of intracellular pH and may also help to increase the pH of extracellular media with a low buffer capacity [5, 8]. Apart from the alkalinisation effect, amino acid decarboxylation can also induce metabolic energy generation through a proton motive force, a fundamental function for bacteria without a high capacity to generate ATP [1, 5, 6, 8].

In general, food products susceptible to containing high levels of histamine are those that are microbiologically spoiled (fresh fish, meat, etc.) or fermented/cured, in which the active bacteria can present an aminogenic capacity [11, 12]. Moreover, foods rich in free histidine may be more susceptible to histamine accumulation, as is the case of scombroid fish species [7, 12]. Endogenous histamine is also found in foods that contain blood or viscera, as well as in some plant products [3, 13].

Storage temperature is one of the most important factors of histamine formation in food [7, 10, 14, 15]. High temperatures (around 25–30°C) have been described by many authors as optimal for most histaminogenic microorganisms, but significant histamine formation has also been reported in refrigerated foods (4–10°C), especially fish [14, 16, 17]. Only ice storage at near 0°C was found to retard histamine formation [17]. In fermentation processes, histamine and other biogenic amines formation is reduced at temperatures lower than 15°C [18]. Other factors, such as pH, formulation (e.g. salting, species, nitrites), starter cultures, technological conservation processes (pasteurisation, high hydrostatic pressures, irradiations) and packaging (vacuum, modified atmospheres), have been extensively studied as potential conditioners of microbial capacity to form histamine [6, 11, 12, 14].

2.2 Hazard characterisation

2.2.1 Physiological role and metabolism of histamine in the organism

Histamine is involved in vital physiological activities in the human body, including neurotransmission, immunomodulation, haematopoiesis, wound healing, circadian rhythms, regulation of cell proliferation, contraction of smooth muscle cells, vasodilatation, increased vascular permeability and mucous secretion, alterations in blood pressure, arrhythmias and the stimulation of gastric acid secretion [15, 19].

Its effects occur through the binding of histamine with four receptors (H1R, H2R, H3R and H4R) located in the target cells of various tissues. Plasma histamine levels between 0.3 and 1.0 ng/mL are considered normal [19].

In humans, two main routes of histamine metabolism are known, involving the enzymes histamine-N-methyltransferase (HNMT) and DAO (**Figure 2**). The product of HNMT-catalysed histamine methylation is N-methylhistamine (MHA), which is subsequently transformed by the monoamine oxidase (MAO) to N-methylimidazole acetaldehyde. The latter is then converted to N-methylimidazoleacetic acid (MIAA) by aldehyde dehydrogenase (ALDH). HNMT, a cytosolic enzyme that metabolises histamine only in the intracellular space, is expressed in almost all tissues, although mainly in the liver and kidney [15, 20]. On the other hand, the oxidative deamination of histamine by DAO leads to imidazole acetaldehyde and then, via ALDH transformation, to imidazole acetic acid (IAA), which is combined with a ribose molecule for its excretion [15, 20]. DAO, which is found mainly in the intestines, placenta and kidneys, is a secretory protein responsible for scavenging extracellular histamine after mediator release [15, 20].

2.2.2 Adverse health effects

2.2.2.1 Histamine intoxication

Histamine intoxication occurs after the ingestion of foods with unusually high amounts of histamine, which overwhelm the metabolic capacity of the organism [1, 21]. Previously known as *scombrotoxicosis, scombroid fish poisoning, pseudoallergic fish poisoning, histamine overdose or mahi-mahi flush* [1, 15], it was initially associated with ingestion of fishes from the *Scombridae* and *Scomberesocidae* families (e.g. mackerel, tuna, albacore and bonito). The first histamine intoxication was reported in sailors some centuries ago, but it was not until 1946 that publications began to describe the relationship between histamine and intoxication symptoms [9, 22, 23]. In the 1980s, the WHO recommended using the term histamine poisoning or intoxication, as the condition could be caused by the consumption of other fish species, as well as other foods [2].

The symptomatology associated with histamine intoxication is closely related to the different physiological actions of histamine in the organism. The main symptoms are neurological, gastrointestinal, dermatological and respiratory, notably rash, erythema, sweating, nausea, vomiting, diarrhoea, a sensation of burning in the mouth, swelling of tongue and face, headache, respiratory distress, palpitations and hypotension [4, 15, 21]. Symptoms are generally mild, appearing 30 minutes after ingestion and disappearing within 24 hours [4, 10, 24]. On rare occasions, they can be more severe and require medical attention [3, 4]. The severity or type of

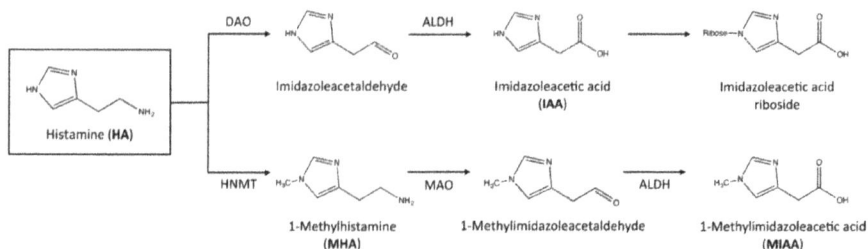

Figure 2.
Metabolic pathways of histamine in the human organism.

reaction depends on the amount of histamine in the plasma. Thus, plasma concentrations higher than the normal level of 1–2 ng/mL result in an increase in gastric secretion and heart rate; 3–5 ng/mL causes headache, flushing, urticaria, pruritus and tachycardia, 6–8 ng/mL a decrease in blood pressure, 7–12 ng/mL bronchospasm and over 100 ng/mL cardiac arrest [19].

The similarity between the described symptomatology and that of allergic reactions can lead to an incorrect diagnosis. Intoxication may be distinguished from a food allergy by taking into account the absence of an allergy history and its occurrence in an outbreak involving more than one patient over a short period of time after the consumption of foods with a high histamine load [9]. For a differential diagnosis, the concentration of serum tryptase measured within 1–2 hours of the onset of symptoms can also be helpful. In food allergy, the activity of serum tryptase increases, whereas in histamine intoxication it should remain within normal physiological values [15, 25]. Moreover, intoxication may be confirmed by elevated histamine levels in the suspected implicated food [1].

2.2.2.2 Histamine intolerance

Histamine intolerance, also known as enteral histaminosis or sensitivity to food histamine, is a disorder in histamine homeostasis mainly due to reduced intestinal degradation by a deficit of DAO activity and the resulting enhanced plasma concentrations [19, 26, 27]. This disorder is not so widely known as histamine intoxication, since the first reference regarding histaminosis or histamine intolerance dates from 1988, and most of the studies have appeared during the last 15 years [28]. DAO enzyme deficiency may have a genetic aetiology. A significant relationship has been found between lower DAO activity and the presence of different single-nucleotide polymorphisms (SNPs) in the AOC1 gene located on chromosome 7 (7q36.1) that encodes this enzyme [29, 30]. The secretion of DAO may also be inhibited by certain pathologies, especially inflammatory bowel diseases, and also by the action of drugs (acetylcysteine, clavulanic acid, metoclopramide, verapamil, etc.) [15, 19]. The role of DAO inhibitor drugs may be significant, as it has been estimated they are being consumed by approximately 20% of the European population [1, 28].

Several clinical studies have linked DAO deficiency with the appearance of gastrointestinal (abdominal pain, diarrhoea and vomiting), dermatological (atopic dermatitis, eczema or chronic urticaria) and/or neurological (headaches) complaints [31–42]. Individuals with histamine intolerance due to DAO deficiency may suffer symptoms similar to those of intoxication, but they appear after a lower histamine intake. The diagnosis is based on the presentation of at least two clinical symptoms, which go into remission when a low-histamine diet is adopted (always after ruling out positive results for food allergy) [19, 42]. Individuals with histamine intolerance can also be identified by determining serum DAO activity, although evidence for the usefulness of this analysis is still scarce and inconclusive [27, 34, 36, 43, 44]. Despite the lack of well-proven data on the incidence of histamine-intolerant individuals with a DAO deficit, it has been estimated as affecting 1% of the population [19].

2.2.3 Dose-response relationship assessment

Although a range of plasma histamine levels have been associated with the onset of different symptoms, as described above, there is no consensus on what quantities of histamine in food are responsible for intoxication outbreaks. Dose-response data from food histamine are scant [24]. According to some studies in healthy volunteers, histamine was found to trigger intoxication symptoms at levels of 75–300 mg when administered with food (fish or non-alcoholic drinks) [1]. When histamine was administered with alcoholic beverages, levels of 0.12–4 mg in wine did not cause significant effects on healthy volunteers, whereas another trial demonstrated the onset of clinical symptoms in 12 out of 40 patients with histamine intolerance following a provocation test with 4 mg of histamine in sparkling wine [1]. Wantke et al. [45] reported the onset of symptoms after the ingestion of 50 µg of histamine in wine (125 mL) in patients with histamine intolerance. On the other hand, when 120 mg of histamine was introduced directly into the duodenum (not transported by food), symptoms appeared in patients diagnosed with chronic urticaria but not in healthy volunteers [1].

The majority of histamine poisoning outbreaks described in the literature occurred after the consumption of high amounts of histamine, mainly in fish [1]. The histamine levels in the foods associated with these outbreaks vary considerably, in the vast majority of cases with values ranging from 100 to 5000 mg/kg [46–53], although amounts of up to 10,000 mg/kg have also been reported [53].

Due to the lack of consensus to establish a threshold toxic dose for histamine intoxication, some authors have proposed some safe levels [3, 4]. Lehane and Olley [54] suggested 30 mg of histamine as a safe dose, calculated from the maximum level of 100 mg/kg of histamine in foods and based on a fish serving of 300 g and a consumer weight of 60 kg. The same authors, however, pointed out that the accuracy of their calculation was limited by an incomplete understanding of histamine intoxication. Later, Rauscher-Gabernig et al. [24] reported that dietary histamine levels in the range of 6–25 mg/meal had no adverse effects. The EFSA expert panel on biological risks proposed 50 mg/person/meal as a safe upper limit of histamine intake for healthy individuals, based on the few studies published to date [1]. This value corresponds with the 50 mg safe threshold advocated for the healthy population by the joint FAO/WHO report on health risks of histamine and other biological amines in fish and derivatives [21].

A safe level of histamine intake for intolerant individuals is not proposed in any of the studies on this disorder. The only recommendation available is from EFSA, which carried out a risk assessment of biogenic amine formation in fermented products and concluded that only foods with histamine levels below detectable limits can be considered safe for intolerant patients [1].

2.2.4 Factors contributing to histamine sensitivity

One of the most important factors affecting sensitivity to dietary histamine is the different histamine-metabolising capacity of each individual. Thus, those with a lower activity of enzymes involved in histamine metabolism (DAO, HNMT, MAO) are more sensitive to suffer the effects after histamine ingestion [1]. An impaired enzymatic activity may have a genetic explanation or be caused by intestinal pathologies or the action of drugs with an inhibitory effect [15, 19, 20]. In this context, the most studied enzyme is DAO. The enhanced sensitivity of women to histamine in certain physiological states, such as in the premenstrual period, is attributed to a reduced DAO activity [55]. Conversely, an increase in DAO production of up to 500-fold has been reported during pregnancy, accompanied by a remission of certain symptoms related to histamine intolerance [56]. Therefore, from the metabolic point of view, there is inter- as well as intra-individual variation in sensitivity.

The toxicity of histamine may be enhanced by dietary components, such as other biogenic amines or alcohol. Putrescine, cadaverine, tyramine and spermidine are biogenic amines usually found together with histamine in food and likewise are DAO substrates. Due to competition for intestinal mucin attachment sites and metabolisation, the ingestion of high quantities of these other amines may potentiate the adverse effects of histamine [1, 10, 15, 54]. This effect has been demonstrated in amines such as putrescine, cadaverine and tyramine, among others, in both in vitro and animal studies [1, 11]. These amines were found to exert an inhibitory effect on histamine metabolism when present at levels 4–5 times higher than that of histamine [11]. This potentiation mechanism could explain why symptoms do not appear when histamine is administered intravenously and yet are triggered when the same amounts of histamine are consumed in foods containing other amines [4].

Alcohol and its metabolite acetaldehyde can also have a potentiating effect. Competition for the ALDH enzyme, which is involved in the metabolism of both alcohol and histamine, results in an accumulation of histamine [1, 12]. The presence of these potentiating factors can thus explain the appearance/absence of symptoms or the variable degrees of reaction among individuals who have consumed foods containing the same amount of histamine.

2.2.5 Outbreaks of histamine intoxication in the European Union

Frequent misdiagnosis and the lack of an adequate and obligatory system for reporting histamine intoxication could account for the limited statistical data on the incidence of this food-borne disease [10, 15]. A total of 386 outbreaks were reported in 2010–2015 by different EU member states according to the EFSA reports on food-borne outbreaks [57]. In 191 of the outbreaks, the food responsible was determined with strong evidence, involving more than 1000 cases of which 107 were hospitalised (**Figure 3**). No deaths due to histamine intoxication were reported during this period. According to these data, there is no clear declining trend in the incidence of histamine intoxication in recent years, in contrast with other types of food poisoning (**Figure 3**). During this period, fish and fishery products were the primary cause of histamine intoxication in the European Union (176 outbreaks), followed by "mixed foods" (six outbreaks, three of which included a dish of tuna), "cheese" (three outbreaks), "buffet meal", "crustaceans, shellfish, mollusc and products thereof", "dairy products other than cheese", "vegetables and juices and other products thereof" (one outbreak each) and "other foods" (two outbreaks) [57].

On the other hand, according to information extracted from the Rapid Alert System for Food and Feed (RASFF), there was a clear rise in notification of histamine intoxication cases linked to tuna consumption during 2014–2017, with a particularly sharp increase in 2017 [57]. In May 2017, Spain and France reported a

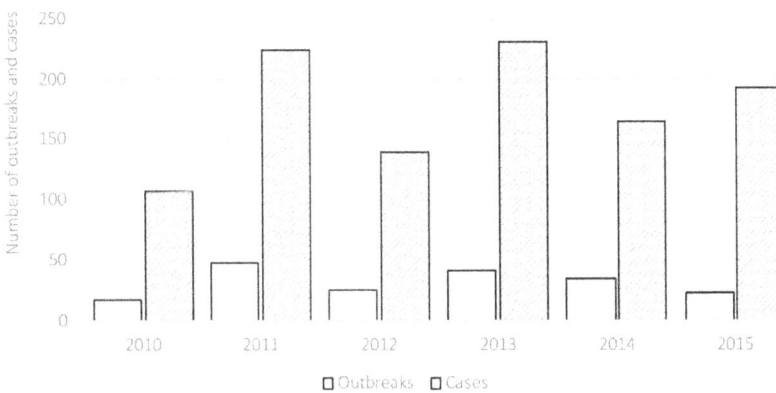

Figure 3.
Assessment of the incidence of histamine intoxication in EU countries during the 2010–2015 [57].

high incidence of histamine intoxication after the consumption of yellowfin tuna from Spain. Additional cases may have arisen in other countries that imported this food product. More than 150 people in Spain and more than 40 in France were affected after consuming tuna that was allegedly treated with a vegetable extract to alter the colour and enhance display freshness [58]. The modification of colour may mask spoilage responsible for the production of histamine and other biogenic amines.

2.3 Exposure assessment

To assess consumer exposure, it is necessary to have data on histamine levels in foods. The overall exposure to dietary histamine is difficult to estimate due to its multiple potential sources and variable concentration. **Figure 4**, which shows the distribution of histamine contents in different foods retailed in Spain, reflects this characteristic high variability, both among different food categories and within the same category.

Fresh fish and fishery products usually do not contain histamine or only low levels [59]. As shown in **Figure 4**, in most of the Spanish retail fish samples reported by Bover-Cid et al. [3], histamine was absent or found in very small quantities (P95 below 5 mg/kg). These data are in agreement with the scientific report published by EFSA based on samples of non-fermented fish and fish products from different European countries, in which only 27% of a total of 6329 samples contained histamine, usually at low concentrations (median of 2.5 mg/kg) [1]. However, a lack of freshness in raw fish and/or hygienically inadequate manufacturing processes of semi-preserved or preserved fish products can lead to markedly high histamine content. An example is the 657 mg/kg of histamine recorded within the Spanish canned fish category (**Figure 4**) or the 8910 mg/kg in fish and fishery products in the EFSA report [1]. Notably, when freshness is lost, in addition to high amounts of histamine, other amines related to the decarboxylase activity of spoilage bacteria, such as cadaverine and putrescine, also frequently accumulate.

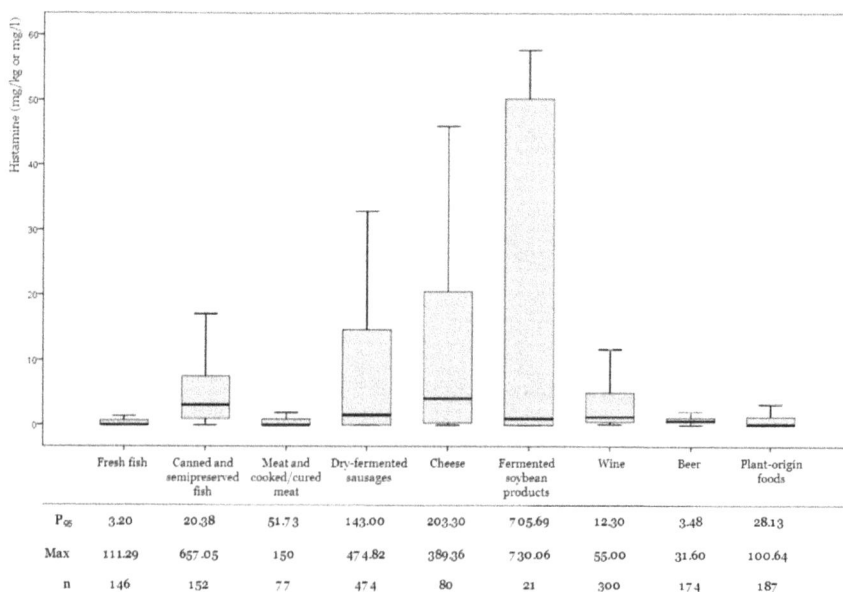

	Fresh fish	Canned and semipreserved fish	Meat and cooked/cured meat	Dry-fermented sausages	Cheese	Fermented soybean products	Wine	Beer	Plant-origin foods
P95	3.20	20.38	51.73	143.00	203.30	705.69	12.30	3.48	28.13
Max	111.29	657.05	150	474.82	389.36	730.06	55.00	31.60	100.64
n	146	152	77	474	80	21	300	174	187

Figure 4.
Histamine distribution (mg/kg or mg/L) in different food products [3].

In fresh, cooked or cured meat, as in fish, no histamine occurrence is expected as long as freshness and a proper hygienic status of the products or manufacturing processes are ensured [1, 3, 60]. In contrast, fermented foods are susceptible to accumulating large amounts of histamine [1, 3, 5, 61]. In this type of foods, the occurrence of histamine depends not only on the hygienic conditions of the raw materials and/or manufacturing processes but also on the aminogenic capacity of the bacteria responsible for the fermentation [11, 12]. As can be seen in **Figure 4**, the presence of histamine in Spanish retail fermented foods is frequently relatively low (85% of samples below 20 mg/kg), but in some cases its levels are notably high, as in cheese (389 mg/kg), fermented sausages (475 mg/kg) or soybean products (730 mg/kg). In fermented beverages (e.g. wine and beer), histamine contents are much lower than those reported for other fermented foods. Notably, together with histamine, tyramine is usually the most frequent and abundant amine in fermented foods, because its formation is closely related to the lactic acid bacteria responsible for the fermentation processes, and its levels can reach 600 mg/kg in cheese, 700 mg/kg in fermented sausages and 1700 mg/kg in fermented vegetables. The occurrence of putrescine and cadaverine is also quite common but at lower and more variable concentrations than tyramine.

Among foods of plant origin, only some vegetables usually show significant levels of histamine, such as spinach, tomato and eggplant [13]. In these products low levels of histamine may have a physiological origin, but undesirable microbial enzymatic activity during storage can lead to the accumulation of high levels [13, 62]. Lavizzari et al. [62] reported a significant increase in histamine levels in different spinach samples stored at refrigeration temperature during 2 weeks. Histamine formation in this type of vegetables was attributed to the activity of some Gram-negative bacteria, mainly belonging to *Enterobacteriaceae* and *Pseudomonadaceae* groups, as their growth is favoured by the relatively high pH of spinach. Asparagus, pumpkin, chard and avocado rarely contain histamine and at very low levels [13]. Other types of frequently consumed foods, such as cereals, milk, yoghurt and eggs, do not show significant contents of histamine or any other biogenic amine [3].

In addition to food content, another fundamental issue when assessing exposure to histamine is the actual consumption of food by the population. Food consumption data need to be as representative as possible, with sources such as the most recent national dietary surveys.

In the exposure assessment performed by EFSA, the 95-percentiles of biogenic amine contents of different European foods and their consumption patterns were combined to provide exposure values in terms of mg/day as an estimation of a high exposure scenario [1]. For histamine, the highest exposure values were obtained for the category "fresh, frozen and canned fish" (8.8–41.4 mg/day) followed by "dry-fermented sausages" (6.4–37.1 mg/day), "cheese" (13–32.1 mg/day) and "fish sauces" (0.4–29.9 mg/day). Likewise, in an assessment of the dietary histamine exposure of Austrian consumers, it was concluded that tuna fish and some fermented products (cheese and sauerkraut) were the top contributors to the total histamine intake [24]. According to this study, a typical meal with fish as a main dish could contribute from 2.3 to 264 mg of histamine. Recently, Latorre-Moratalla et al. [63] carried out an assessment of Spanish consumer exposure to histamine derived from the consumption of dry-fermented sausages, concluding that in 95% of cases, it was lower than 6.8 mg/meal.

The larger serving size of fish products, together with the extremely high histamine contents arising from hygienic defects in their conservation or manufacturing processes, could explain why these foods contribute more to histamine exposure than others, such as cheese or dry-fermented sausages, which a priori have higher average histamine contents. This could also explain why fish and fishery products are the predominant cause of histamine intoxication.

2.4 Risk characterisation

Performing an adequate quantitative risk characterisation of histamine exposure is hampered by the lack of dose-response data. However, a qualitative approach, taking into account the limited data available, suggests that the risk of suffering histamine intoxication is relatively low, since exposure exceeds the safety limit on very few occasions.

According to the qualitative risk characterisation performed by EFSA, exposure to histamine (95-percentile value) in fermented foods does not go beyond the safe threshold of 50 mg/meal/person [1]. However, it is stressed that this upper limit may be exceeded by the consumption of more than one food item with high histamine content during the same meal. Likewise, Latorre-Moratalla et al. [63] concluded that the risk of suffering acute effects related to histamine intake after the consumption of dry-fermented sausages in Spain is very low, since the exposure levels rarely exceed the safe threshold. It should be noted that all available studies have been carried out on specific food groups and to date none have dealt with the full range of histamine-containing foods.

The risk for the histamine-intolerant population is higher, because even small amounts of histamine may trigger adverse effects [1, 63]. No studies have been carried out to evaluate this risk.

3. Risk management

From the perspective of risk management, decision-makers or the food industry should consider different actions to reduce consumer exposure to dietary histamine. For example, regulatory measures or strategies to prevent, reduce or eliminate the presence of histamine and other amines in foods could be implemented.

From a legal perspective, tolerable limits of histamine have been fixed by different countries only for fish and fishery products. The European Union through Regulation (EC) No. 2073/2005 on microbiological criteria has established that in fishery products from species with a naturally high histidine content (particularly fish from the *Scombridae, Clupeidae, Engraulidae, Coryphaenidae, Pomatomidae* and *Scomberesocidae* families), using a sampling plan of nine samples, the mean histamine value must be less than 100 mg/kg, no sample should exceed 200 mg/kg and among nine samples only two can have a histamine content between 100 and 200 mg/kg [64]. In fishery products obtained from species with a high histidine content and subjected to an enzymatic maturation treatment in brine, the maximum average value allowed is 200 mg/kg, the maximum individual value is 400 mg/kg and there can only be two samples between these two values in a sampling plan of nine samples [64, 65]. The Food and Drug Administration (FDA) of the United States, using a sampling plan of 18 samples, establishes a tolerable maximum level of 50 mg/kg of histamine for tuna and mahi-mahi or between 50 and 500 mg/kg of histamine for other fish species, with just one sample higher than these values [66]. In Canada, Switzerland and Brazil, the maximum limit allowed in fish and fishery products is 100 mg/kg and in Australia and New Zealand 200 mg/kg [7].

As histamine and other biogenic amines are a food safety concern, the food industry needs to consider improving control strategies. Available knowledge about biogenic amine formation in certain foods has made it possible to design measures to prevent or at least reduce their accumulation during manufacture and storage.

A key strategy is to guarantee and improve the hygienic quality of raw materials and manufacturing processes. Since contaminant microorganisms are responsible for biogenic amine formation in many products, food quality and safety management based on hazard analysis and critical control points (HACCP) are essential [4, 7]. The time/temperature binomial is the most critical and determinant risk factor in biogenic amine formation in most fresh and lightly preserved meat and seafood products, as well as in raw materials (of animal and plant origin) used in the preparation of cooked, ripened or fermented products [3, 7, 11]. Predictive models of biogenic amine formation in perishable products as a function of time and temperature could be used to avoid hazardous storage conditions [11, 67]. This approach has already been implemented to reduce histamine accumulation in seafood, which is particularly associated with histamine formation by *Morganella psychrotolerans* and *M. morganii* [67]. Other factors that inhibit or reduce aminogenic activity, such as the packaging atmosphere and the addition of salt and other preservatives, should be taken into account for lightly preserved fish, meat and vegetables [6, 11].

In the case of fermented foods or beverages, the hygienic quality of raw materials may be enhanced by decreasing the microbial load through pasteurisation, a common practice in the cheese-making industry [5]. However, in fermented meat products, high temperature causes detrimental changes in the raw materials, thus rendering the conventional heat treatments unsuitable. The application of high-pressure treatments to raw materials (milk and meat) could be an effective strategy to improve their hygienic status, thereby reducing the biogenic amine accumulation without significant alteration of sensory properties [11, 68].

Techniques that avoid biogenic amine formation should also be implemented in fermented food products. For example, the use of a suitable formulation (adjusting the type and amount of fermentable sugar, spices and preservatives) and well-established technological parameters (temperature and relative humidity) enables a quick and accurate selection of desired fermentative microbiota, which limits the action of contaminant microorganisms, including amine formation [11, 18]. Moreover, it has been widely demonstrated that the selection of microorganisms without aminogenic activity for the starter cultures constitutes an effective control measure against biogenic amine accumulation in fermented products [6, 11, 18, 61]. For cases where these strategies are not sufficiently effective, a new approach currently being explored is the application of starter cultures with amino-oxidase activity, which are able to degrade previously formed amines [11, 61].

A general control strategy for wine that includes many of these approaches is the so-called low-histamine technology, which is based on the guaranteed hygienic quality of the raw materials, the addition of selected starter cultures and the use of specific production techniques that inhibit histamine formation [3]. The current challenge for the food industry is to extend this technology to other biogenic amines and other products. The successful implementation of "low biogenic amine technologies" will enable food manufacturers to produce safe and high-quality amine-free food.

4. Conclusions

Although the health problems associated with histamine consumption are well known, there are some uncertainties, especially regarding the threshold level of this amine that may trigger the symptoms. Histamine concentrations reported as responsible for intoxication outbreaks are extremely variable. Thus, despite the association of adverse effects with the ingestion of high levels of histamine, lower levels can also cause symptoms in sensitive individuals with a genetic or acquired impairment in the enzymatic degradation of histamine (the histamine-intolerant population). The difficulty in establishing a specific toxic dose is also due to the presence of other biogenic amines in the implicated foods, which can potentiate the adverse effects of histamine.

The few qualitative risk assessments of histamine performed to date all indicate that the current risk of suffering histamine intoxication is low. However, in the population with histamine intolerance, the risk of suffering symptoms derived from the intake of histamine would be much higher. In fact, according to the risk assessment performed by EFSA, only levels below the detectable limit can be considered as safe. Therefore, for these individuals, the main strategy to avoid histamine-related health problems is to follow a diet that excludes foods rich in this or other amines, such as putrescine and cadaverine, which can potentiate the adverse effects of histamine.

According to current EU data, the consumption of spoiled fish and fishery products is the main cause of histamine intoxication outbreaks. Although this type of food usually has a low or negligible histamine content, a lack of freshness and poor hygienic conditions may result in the accumulation of high levels and trigger an outbreak. In contrast, whereas fermented foods often have higher histamine contents than fish and fish derivatives, their serving size tends to be lower, which could explain why they are generally less implicated in the outbreaks. Nevertheless, even when the risk of intoxication is low, the accumulation of high levels of histamine in fermented foods may indicate poor hygienic quality and is an argument for extending the legislative criteria to foods other than fish.

The lack of consensus on what quantities of dietary histamine produce intoxication can be explained by the coexistence of other biogenic amines in the same foods, as well as inter- and intra-individual variability in histamine metabolisation. The impaired ability to metabolise ingested histamine has led to the description of a relatively new disorder, that of histamine intolerance due to DAO deficiency.

A current challenge for the food industry is to offer products with minimal biogenic amine levels, and it may be recommendable to declare the presence or absence of histamine in food labelling. Private initiatives have already begun to include the message "without histamine" in products as a hallmark of quality. Such labelling could help those with histamine intolerance to make a more informed selection of suitable foods.

Acknowledgements

Sònia Sánchez-Pérez is a recipient of a doctoral fellowship from the University of Barcelona (APIF2018).

Conflict of interest

The authors declare no conflict of interest.

Author details

Oriol Comas-Basté[1,2,3], Maria Luz Latorre-Moratalla[1,2,3], Sònia Sánchez-Pérez[1,2,3], Maria Teresa Veciana-Nogués[1,2,3] and Maria del Carmen Vidal-Carou[1,2,3*]

1 Department of Nutrition, Food Sciences and Gastronomy, Faculty of Pharmacy and Food Sciences, University of Barcelona (UB), Santa Coloma de Gramenet, Spain

2 Research Institute of Nutrition and Food Safety of the University of Barcelona (INSA·UB), Santa Coloma de Gramenet, Spain

3 Catalonian Reference Network on Food Technology (XaRTA), Barcelona, Spain

*Address all correspondence to: mcvidal@ub.edu

IntechOpen

References

[1] EFSA Panel on Biological Hazards (BIOHAZ). Scientific opinion on risk based control of biogenic amines formation in fermented foods. EFSA Journal. 2011;**9**:2393. DOI: 10.2903/j.efsa.2011.2393

[2] Taylor SL. Histamine Poisoning Associated with Fish, Cheese and Other Foods. Report VPH/FOS/85.1. Geneve: World Health Organisation Press; 1985. pp. 1-48

[3] Bover-Cid S, Latorre-Moratalla ML, Veciana-Nogués MT, Vidal-Carou MC. Processing contaminants: Biogenic amines. In: Motarjemi Y, Moy GG, Todd ECD, editors. Encyclopedia of Food Safety. Vol. 2. Waltham: Elsevier Inc; 2014. pp. 381-391. DOI: 10.1016/B978-0-12-378612-8.00216-X

[4] Hungerfort JM. Scombroid poisoning: A review. Toxicon. 2010;**56**:231-243. DOI: 10.1016/j.toxicon.2010.02.006

[5] Linares DM, Martín MC, Ladero V, Alvarez MA, Fernández M. Biogenic amines in dairy products. Critical Reviews in Food Science and Nutrition. 2011;**51**:691-703. DOI: 10.1080/10408398.2011.582813

[6] Gardini F, Özogul Y, Suzzi G, Tabanelli G, Özogul F. Technological factors affecting biogenic amine content in foods: A review. Frontiers in Microbiology. 2016;**7**:1218. DOI: 10.3389/fmicb.2016.01218

[7] Biji KB, Ravishankar CN, Venkateswarlu R, Mohan CO, Srinivasa Gopal TK. Biogenic amines in seafood: A review. Journal of Food Science and Technology. 2016;**53**:2210-2218. DOI: 10.1007/s13197-016-2224-x

[8] Russo P, Spano G, Arena MP, Capozzi V, Fiocco D, Grieco F, et al. Are consumers aware of the risks related to biogenic amines in food? In: Mendez-Villas A, editor. Current Research, Technology and Education Topics in Applied Microbiology and Microbial Biotechnology. Badajoz: Formatex Research Center; 2010. pp. 1087-1095

[9] Ladero V, Calles M, Fernandez M, Alvarez MA. Toxicological effects of dietary biogenic amines. Current Nutrition & Food Science. 2010;**6**:145-156. DOI: 10.2174/157340110791233256

[10] Visciano P, Schirone M, Tofalo R, Suzzi G. Histamine poisoning and control measures in fish and fishery products. Frontiers in Microbiology. 2014;**5**:500. DOI: 10.3389/fmicb.2014.00500

[11] Naila A, Flint S, Fletcher G, Bremer P, Meerdink G. Control of biogenic amines in food-existing and emerging approaches. Journal of Food Science. 2010;**75**:139-150. DOI: 10.1111/j.1750-3841.2010.01774.x

[12] Vidal-Carou MC, Veciana-Nogués MT, Latorre-Moratalla ML, Bover-Cid S. Biogenic amines: Risks and control. In: Toldrá F, Hui YH, Astiasarán I, Sebranek JG, Talon R, editors. Handbook of Fermented Meat and Poultry. 2nd ed. Oxford: Wiley-Blackwell; 2014. pp. 413-428. DOI: 10.1002/9781118522653.ch47

[13] Sánchez-Pérez S, Comas-Basté O, Rabell-González J, Veciana-Nogués MT, Latorre-Moratalla ML, Vidal-Carou MC. Biogenic amines in plant-origin foods: Are they frequently underestimated in low-histamine diets? Food. 2018;**14**:7-12. DOI: 10.3390/foods7120205

[14] Chong CY, Abu Bakar F, Russly AR, Jamilah B, Mahyudin NA. The effects of food processing on biogenic amines formation. International Food Research Journal. 2011;**18**:867-876

[15] Kovacova-Hanuskova E, Buday T, Gavliakova S, Plevkova J. Histamine, histamine intoxication and intolerance. Allergologia et Immunopathologia. 2015;**43**:498-506. DOI: 10.1016/j.aller.2015.05.001

[16] Naila A, Flint S, Fletcher G, Bremer P, Meerdink G. Histamine stability in Rihaakuru at −80, 4 and 30°C. Food Chemistry. 2012;**135**:1226-1229. DOI: 10.1016/j.foodchem.2012.05.066

[17] Jiang QQ, Dai ZY, Zhou T, Wu JJ, Bu JZ, Zheng TL. Histamine production and bacterial growth in mackerel (*Pneumatophorus japonicus*) during storage. Journal of Food Biochemistry. 2013;**37**:246-253. DOI: 10.1111/jfbc.12021

[18] Latorre-Moratalla ML, Bover-Cid S, Veciana-Nogués MT, Vidal-Carou MC. Control of biogenic amines in fermented sausages: Role of starter cultures. Frontiers in Microbiology. 2012;**3**:169. DOI: 10.3389/fmicb.2012.00169

[19] Maintz L, Novak N. Histamine and histamine intolerance. The American Journal of Clinical Nutrition. 2007;**85**:1185-1196. DOI: 10.1093/ajcn/85.5.1185

[20] Comas-Basté O, Latorre-Moratalla ML, Bernacchia R, Veciana-Nogués MT, Vidal-Carou MC. New approach for the diagnosis of histamine intolerance based on the determination of histamine and methylhistamine in urine. Journal of Pharmaceutical and Biomedical Analysis. 2017;**145**:379-385. DOI: 10.1016/j.jpba.2017.06.029

[21] World Health Organization & Food and Agriculture Organization of the United Nations. Joint FAO/WHO Expert Meeting on the Public Health Risks of Histamine and Other Biogenic Amines from Fish and Fishery Products: Meeting Report. World Health Organization; 2013

[22] Legroux R, Levaditi JC, Segond L. Methode de mise en evidence de l'histamine dans les aliments causes d'intoxications collectives a l'aide de l'inoculation au cobaye. Comptes Rendus Biologies. 1946;**140**:863-864

[23] Legroux R, Levaditi JC, Bouidin G, Bovet D. Intoxications histaminiques collectives consecutives a l'ingestion de thon frais. La Presse Médicale. 1964;**54**:545-546

[24] Rauscher-Gabernig E, Grossgut R, Bauer F, Paulsen P. Assessment of alimentary histamine exposure of consumers in Austria and development of tolerable levels in typical foods. Food Control. 2009;**20**:423-429. DOI: 10.1016/j.foodcont.2008.07.011

[25] Ricci G, Zannoni M, Cigolini D, Caroselli C, Codogni R, Caruso B, et al. Tryptase serum level as a possible indicator of scombroid syndrome. Clinical Toxicology. 2010;**48**:203-206. DOI: 10.3109/15563651003649177

[26] Schwelberger HG. Histamine intolerance: A metabolic disease? Inflammation Research. 2010;**59**:S219-S221. DOI: 10.1007/s00011-009-0134-3

[27] Reese I, Ballmer-Weber B, Beyer K, Fuchs T, Kleine-Tebbe J, Klimek L, et al. German guideline for the management of adverse reactions to ingested histamine: Guideline of the German society for allergology and clinical immunology (DGAKI), the German society for pediatric allergology and environmental medicine (GPA), the German association of allergologists (AeDA), and the Swiss society for allergology and immunology (SGAI). Allergo Journal International. 2017;**26**:72-79. DOI: 10.1007/s40629-017-0011-5

[28] Sattler J, Häfner D, Klotter HJ, Lorenz W, Wagner PK. Food-induced histaminosis as an epidemiological

problem: Plasma histamine elevation and haemodynamic alterations after oral histamine administration and blockade of diamine oxidase (DAO). Agents and Actions. 1988;**23**:361-365

[29] Maintz L, Yu CF, Rodríguez E, Baurecht H, Bieber T, Illig T, et al. Association of single nucleotide polymorphisms in the diamine oxidase gene with diamine oxidase serum activities. Allergy. 2011;**66**:893-902. DOI: 10.1111/j.1398-9995.2011.02548.x

[30] García-Martín E, Martínez C, Serrador M, Alonso-Navarro H, Ayuso P, Navacerrada F, et al. Diamine oxidase rs10156191 and rs2052129 variants are associated with the risk for migraine. Headache. 2015;**55**:276-286. DOI: 10.1111/head.12493

[31] Steinbrecher I, Jarisch R. Histamine and headache. Allergologie. 2005;**28**:85-91

[32] Worm M, Fielder E, Döle AS, Schink T, Hemmer W, Jarisch R, et al. Exogenous histamine aggravates eczema in a subgroup of patients with atopic dermatitis. Acta Dermato-Venereologica. 2009;**89**:52-56. DOI: 10.2340/ 00015555-0565

[33] Honzawa Y, Nakase H, Matsuura M, Chiba T. Clinical significance of serum diamine oxidase activity in inflammatory bowel disease: Importance of evaluation of small intestinal permeability. Inflammatory Bowel Diseases. 2011;**17**:E23-EE5. DOI: 10.1002/ibd.21588

[34] Mušič E, Korošec P, Šilar M, Adamič K, Košnik M, Rijavec M. Serum diamine oxidase activity as a diagnostic test for histamine intolerance. Wiener Klinische Wochenschrift. 2013;**12**:239-243. DOI: 10.1007/s00508-013-0354-y

[35] Rosell-Camps A, Zibetti S, Pérez-Esteban G, Vila-Vidal M, Ferrés Ramis L, García-Teresa-García

E. Intolerancia a la histamina como causa de síntomas digestivos crónicos en pacientes pediátricos. Revista Española de Enfermedades Digestivas. 2013;**105**:201-207. DOI: 10.4321/ S1130-01082013000400004

[36] Manzotti G, Breda D, Gioacchino M, Burastero SE. Serum diamine oxidase activity in patients with histamine intolerance. International Journal of Immunopathology and Pharmacology. 2015;**29**:105-111. DOI: 10.1177/0394632015617170

[37] Hoffmann M, Gruber E, Deutschmann A, Jahnel J, Hauer A. Histamine intolerance in children with chronic abdominal pain. Archives of Disease in Childhood. 2016;**98**:832-833. DOI: 10.1136/ archdischild-2013-305024

[38] Wagner N, Dirk D, Peveling-Oberhag A, Reese I, Rady-Pizarro U, Mitzel H, et al. A popular myth—Low-histamine diet improves chronic spontaneous urticaria—Fact or fiction? Journal of the European Academy of Dermatology and Venereology. 2017;**31**:650-655. DOI: 10.1111/jdv.13966

[39] Kacik J, Wróblewska B, Lewicki S, Zdanowski R, Kalicki B. Serum diamine oxidase in pseudoallergy in the pediatric population. Advances in Experimental Medicine and Biology. 2018;**1039**:35-44. DOI: 10.1007/5584_2017_81

[40] Pinzer TC, Tietz E, Waldmann E, Schink M, Neurath MF, Zopf Y. Circadian profiling reveals higher histamine plasma levels and lower diamine oxidase serum activities in 24% of patients with suspected histamine intolerance compared to food allergy and controls. Allergy. 2018;**73**:949-957. DOI: 10.1111/all.13361

[41] Son JH, Chung BY, Kim HO, Park CW. A histamine-free diet is helpful for treatment of adult patients with chronic spontaneous urticaria. Annals

of Dermatology. 2018;**30**:164-172. DOI: 10.5021/ad.2018.30.2.164

[42] Izquierdo-Casas J, Comas-Basté O, Latorre-Moratalla ML, Lorente-Gascón M, Duelo A, Vidal-Carou MC, et al. Low serum diamine oxidase (DAO) activity levels in patients with migraine. Journal of Physiology and Biochemistry. 2018;**74**:93-99. DOI: 10.1007/s13105-017-0571-3

[43] Töndury B, Wüthrich B, Schmid-Grendelmeier P, Seifert B, Ballmer-Weber BK. Histamine intolerance: Is the determination of diamineoxidase activity in the serum useful in routine clinical practice? Allergologie. 2008;**31**:350-356. DOI: 10.5167/uzh-5336

[44] Kofler H, Aberer W, Beibl M, Hawranek T, Klein G, Reider N, et al. Diamine oxidase (DAO) serum activity: Not a useful marker for diagnosis of histamine intolerance. Allergologie. 2009;**32**:105-109. DOI: 10.5414/ALP32105

[45] Wantke F, Götz M, Jarisch R. The red wine provocation test: Intolerance to histamine as a model for food intolerance. Allergy Proceedings. 1994;**15**:27-32

[46] Dalgaard P, Emborg J, Kjolby A, Sorensen N, Ballin N. Histamine and biogenic amines: Formation and importance in seafood. In: Borrensen T, editor. Improving Seafood Products for the Consumer. Cambridge, UK: Woodhead Publishing Ltd; 2008. pp. 292-324. DOI: 10.1533/9781845694586.3.292

[47] NSW Food Authority. Presence of Histamine in Anchovies. NSW/FA/FI079/1007. 2011

[48] Demoncheaux JP, Michel R, Mazenot C, Duflos G. A large outbreak of scombroid fish poisoning associated with eating yellowfin tuna (*Thunnus albacares*) at a military mass catering

in Dakar, Senegal. Epidemiology and Infection. 2012;**140**:1008-1012. DOI: 10.1017/S0950268811001701

[49] Tortorella V, Masciari P, Pezzi M, Mola A, Tiburzi SP, Zinzi MC, et al. Histamine poisoning from ingestion of fish or scombroid syndrome. Case Reports in Emergency Medicine. 2014;**2014**:4. DOI: 10.1155/2014/482531

[50] Fariñas-Cabrero MA, Berbel-Hernández C, Allué-Tango M, Díez-Hillera M, Herrero-Marcos JA. Outbreak due to butterfish consumption: Keriorrhea and histamine poisoning. Revista Española de Salud Pública. 2015;**89**:99-105. DOI: 10.4321/S1135-57272015000100011

[51] Petrovic J, Babić J, Jaksic S, Kartalovic B, Ljubojevic D, Cirkovic M. Fish product-borne histamine intoxication outbreak and survey of imported fish and fish products in Serbia. Journal of Food Protection. 2016;**79**:90-94. DOI: 10.4315/0362-028X.JFP-15-190

[52] Lee YC, Kung HF, Wu CH, Hsu HM, Chen HC, Huang TC, et al. Determination of histamine in milkfish stick implicated in food-borne poisoning. Journal of Food and Drug Analysis. 2016;**24**:63-71. DOI: 10.1016/j.jfda.2015.06.009

[53] Colombo FM, Cattaneo P, Confalonieri E, Bernardi C. Histamine food poisonings: A systematic review and meta-analysis. Critical Reviews in Food Science and Nutrition. 2018;**58**:1131-1151. DOI: 10.1080/10408398.2016.1242476

[54] Lehane L, Olley J. Histamine fish poisoning revisited. International Journal of Food Microbiology. 2000;**58**:1-37. DOI: 10.1016/S0168-1605(00)00296-8

[55] Hamada Y, Shinohara Y, Yano M, Yamamoto M, Yoshio M, Satake K,

et al. Effect of the menstrual cycle on serum diamine oxidase levels in healthy women. Clinical Biochemistry. 2013;**46**:99-102. DOI: 10.1016/j.clinbiochem.2012.10.013

[56] Maintz L, Schwarzer V, Bieber T, van der Ven K, Novak N. Effects of histamine and diamine oxidase activities on pregnancy: A critical review. Human Reproduction Update. 2008;**14**:485-495. DOI: 10.1093/humupd/dmn014

[57] European Food Safety Authority. Assessment of the incidents of histamine intoxication in some EU countries. EFSA Supporting Publications. 2017;**EN-1301**:37. DOI: 10.2903/sp.efsa.2017.EN-1301

[58] Whitworth JJ. Spain Takes Measures to Address Histamine in Tuna RAFFS Alerts [Internet] 2018. Available from: https://www.foodnavigator.com/Article/2018/06/12/Audit-of-Spanish-tuna-sector-after-histamine-RASFF-notifications [Accessed: May 12, 2018]

[59] Baixas-Nogueras S, Bover-Cid S, Veciana-Nogués MT, Vidal-Carou MC. Effect of gutting on microbial loads, sensory properties, and volatile and biogenic amine contents of European hake (*Merluccius merluccius* var. mediterraneus) stored in ice. Journal of Food Protection. 2009;**72**:1671-1676. DOI: 10.4315/0362-028X-72.8.1671

[60] Latorre-Moratalla ML, Veciana-Nogués MT, Bover-Cid S, Garriga M, Aymerich T, Zanardi E, et al. Biogenic amines in traditional fermented sausages produced in selected European countries. Food Chemistry. 2008;**107**:912-921. DOI: 10.1016/j.foodchem.2007.08.046

[61] Alvarez M, Moreno-Arribas V. The problem of biogenic amines in fermented foods and the use of potential biogenic amine-degrading microorganisms as a solution. Trends

in Food Science and Technology. 2014;**39**:146-155. DOI: 10.1016/j.tifs.2014.07.007

[62] Lavizzari T, Veciana-Nogués MT, Weingart O, Bover-Cid S, Mariné-Font A, Vidal-Carou MC. Occurrence of biogenic amines and polyamines in spinach and changes during storage under refrigeration. Journal of Agricultural and Food Chemistry. 2007;**55**:9514-9519. DOI: 10.1021/jf071307

[63] Latorre-Moratalla ML, Comas-Basté O, Bover-Cid S, Vidal-Carou MC. Tyramine and histamine risk assessment related to consumption of dry fermented sausages by the Spanish population. Food and Chemical Toxicology. 2017;**99**:78-85. DOI: 10.1016/j.fct.2016.11.011

[64] Regulation 2073/2005/EC. Microbiological criteria for foodstuffs. Official Journal of the European Union. 2005;**338**:1-26

[65] Regulation 1019/2013/EU. Amending annex I to regulation EC No 2073/2005 as regards histamine in fishery products. Official Journal of the European Union. 2013;**282**:46-47

[66] Food and Drug Administration (FDA). Fish and Fishery Products Hazards and Controls Guidance. 4th ed. Washington: United States Department of Health and Human Services; 2011. p. 468

[67] Emborg J, Dalgaard P. Growth, inactivation and histamine formation of *Morganella psychrotolerans* and *Morganella morganii*—Development and evaluation of predictive models. International Journal of Food Microbiology. 2008;**128**:234-243. DOI: 10.1016/j.ijfoodmicro.2008.08.015

[68] Calzada J, Del Olmo A, Picón A, Gaya P, Nuñez M. Reducing biogenic-amine-producing bacteria,

decarboxylase activity, and biogenic amines in raw milk cheese by high-pressure treatments. Applied and Environmental Microbiology. 2013;**79**:1277-1283. DOI: 10.1128/AEM.03368-12

Pharmacological Properties of Biogenic Amines

Chapter 5

Biochemical and Pharmacological Properties of Biogenic Amines

Dincer Erdag, Oguz Merhan and Baris Yildiz

Abstract

Biogenic amines are low molecular weight organic nitrogen compounds. They are formed by the decarboxylation of amino acids or by amination and transamination of aldehydes and ketones during normal metabolic processes in living cells and therefore are ubiquitous in animals, plants, microorganisms, and humans. In food and beverages, they are formed by the enzymes of raw materials or are generated by microbial decarboxylation of amino acids. The structure of a biogenic amine can be aromatic and heterocyclic amines (histamine, tryptamine, tyramine, phenylethylamine, and serotonin); aliphatic di-, tri-, and polyamines (putrescine, cadaverine, spermine, spermidine, and agmatine); and aliphatic volatile amines (ethylamine, methylamine, isopentylamine, and ethanolamine). Many of them possess a strong pharmacologic effect, and others are important as precursors of hormones and components of coenzymes. The biogenic amine intoxication leads to toxicological risks and health hazards that trigger psychoactive, vasoactive, and hypertensive effects resulting from consumption of high amounts of biogenic amines in foods. The toxicological effects of biogenic amines increase when the mono- and diaminoxidase enzymes are deficient or drugs that inhibit these enzymes (pain reliever, stress, and depression drugs) are used. In this chapter, biosynthesis of biogenic amines, their toxic effects as well as their physiological functions, and their effect on health will be described.

Keywords: biochemistry, biogenic amines, health, pharmacology, toxicity

1. Introduction

Biogenic amines found in animals, plants, microorganisms, and humans are formed by the decarboxylation of amino acids or amination and transamination of aldehydes and ketones during the standard metabolic processes.

Biogenic amines, having several critical biological roles in the body, have essential physiological functions such as the regulation of growth and blood pressure and control of the nerve conduction. Besides, they are required in the immunologic system of intestines and in maintaining the activity of the standard metabolic functions, and when taking the nourishment in high concentrations, they cause disorders in nervous, respiratory, and cardiovascular systems and allergic reactions as well. In this chapter, biosynthesis of biogenic amines, their toxic effects as well as their physiological functions, and their effect on health will be presented.

2. Biogenic amines

2.1 Decarboxylation of amino acids

CO_2 and biogenic amine occur as a result of the enzymatic reaction catalyzed by pyridoxal phosphate to decarboxylate the amino acid (**Figure 1**) [1]. Biogenic amines are biologically active molecules, as they are formed by decarboxylation of amino acids or amination and transamination of aldehydes and ketones during standard metabolic processes [2]. Biogenic amines take charge of the proliferation and differentiation of cells and their metabolism by entering into the structure of hormones, cobalamin (vitamin and aminoacetone), and coenzyme A in the body [3]. They have importance regarding the environment by causing water pollution, as their formations pertain to the amino acid and microorganisms [3, 4]. Biogenic amines may cause intoxications when taken in high amounts [5].

2.2 Classification of the biogenic amines

Biogenic amines are organic nitrogen compounds having a low molecular weight [5, 6]. Their chemical structure can be classified as (i) aromatic and heterocyclic (histamine, tryptamine, tyramine, phenylethylamine, and serotonin); (ii) aliphatic di-, tri-, and polyamines (putrescine, cadaverine, spermine, spermidine, and agmatine); and (iii) aliphatic volatile amines (ethylamine, methylamine, isopentylamine, and ethanolamine) (**Figure 2**) [7, 8]. Besides, their amine group classifications include (i) monoamine (phenylethylamine, tyramine, methylamine, ethylamine, isopentylamine, and ethanolamine), (ii) diamine (histamine, tryptamine, serotonin, putrescine, and cadaverine), and (iii) polyamine (spermine, spermidine, and agmatine) [7–9].

2.3 Biosynthesis and functions

Biogenic amines generally occur as a result of free amino acid decarboxylations with the microbial enzymes. Amino acid decarboxylation happens by removal of the α-carboxyl group [10]. Their occurrences are as below: histamine from histidine amino acid, tyramine from tyrosine amino acid, tryptamine and serotonin from tryptophan amino acid, phenylethylamine from phenylalanine amino acid, putrescine from ornithine amino acid, cadaverine from lysine amino acid, and agmatine from arginine amino acid (**Figure 3**) [11–13].

Biogenic amines play an essential role in cell membrane stabilization, immune functions, and prevention of chronic diseases, as they participate in the nucleic acid and protein synthesis [14]. Besides, they are compounds created as the growth regulation (spermine, spermidine, and cadaverine), neural transmission (serotonin), and inflammation mediators (histamine and tyramine) [6, 15].

Histamine, a standard component of the body, consists of histidine amino acid as a result of histidine decarboxylase activity depending on pyridoxal phosphate

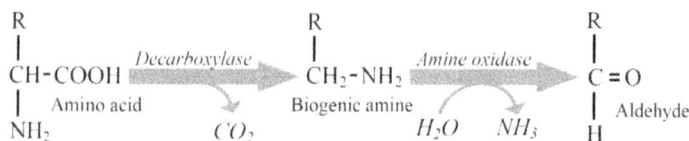

Figure 1.
Decarboxylation of amino acids.

(**Figure 3**) [16]. Histamine distribution and concentration found in the tissues of all vertebrates are very unsteady [17, 18]. Histamine takes charge of some functions related to balancing the body temperature and regulating the stomach volume, stomach pH, and cerebral activities [19] as it participates in the essential functions such as neurotransmission and vascular permeability [20, 21]. However, it also plays a role in starting the allergic reactions [22, 23].

Tryptamine consists of tryptophan amino acid as a result of the aromatic L-amino acid decarboxylase activity (**Figure 3**) [24, 25]. Tryptamine is a monoamine alkaloid found in plants, fungi, and animals [26]. Tryptamine, found in trace amounts in mammalian brains, increases blood pressure [10, 27] as well as plays a role as a neurotransmitter or neuromodulator [26].

The amino acid of phenylalanine synthesizes phenylethylamine through the aromatic L-amino acid decarboxylase in humans, some fungi, and bacteria as well as several plants and animal species (**Figure 3**) [28–30]. It functions as a neurotransmitter in the human central nervous system [31, 32].

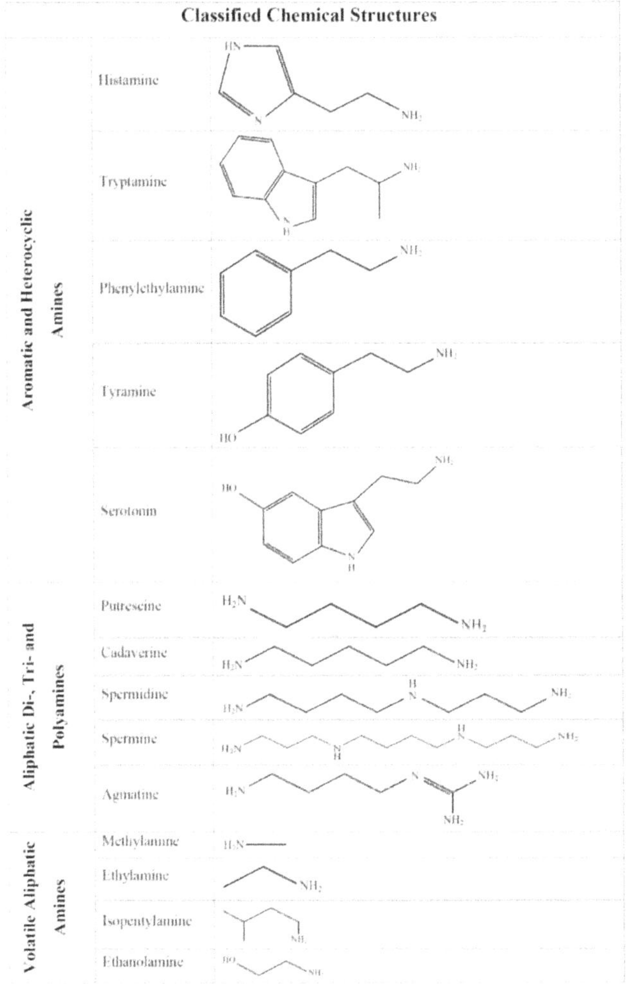

Figure 2.
Classification of biogenic amines according to their chemical structures.

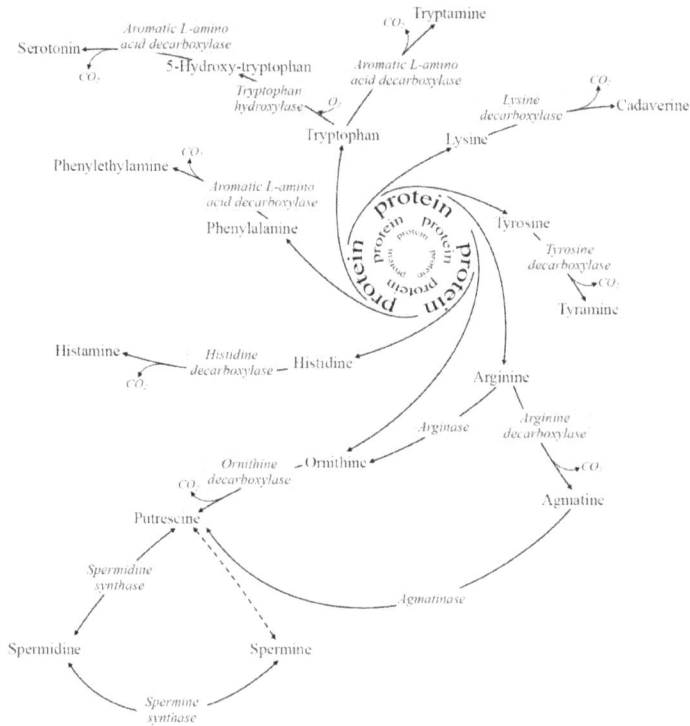

Figure 3.
Formation mechanism of biogenic amines.

Tyramine, consisting of tyrosine amino acid as a result of tyrosine decarboxylase activity, is generally found in low amounts (**Figure 3**) [33–36]. Tyramine leads to several physiological reactions such as blood pressure increase, vasoconstriction [37], tyramine active noradrenalin secretion, etc., as the sympathetic nervous system controls several functions of the body [38, 39]. Tyramine, stored in the neurons, causes the increase in the tear, salivation and respiratory as well as mydriasis [39].

Tryptophan synthesizes serotonin as a result of tryptophan hydroxylase and aromatic L-amino acid decarboxylase enzyme activities (**Figure 3**) [11, 40]. Serotonin, one of the crucial neurotransmitters of the central nervous system, plays a role in plenty of critical physiological mechanisms such as sleep, mood disorders, appetite regulation, sexual behavior, cerebral blood flow regulation, and blood-brain barrier permeability [41, 42].

Putrescine consists of ornithine amino acid as a result of ornithine decarboxylase activity. Besides, it may be synthesized by arginine through the agmatine and carbamoylputrescine (**Figure 3**) [12, 39, 43, 44]. Putrescine, produced by bacteria and fungi, contributes to the cell growth, cell division, and tumorigenesis [45, 46] as it is the preliminary substance of spermidine and spermine [12, 47].

Cadaverine, synthesized by lysine as a result of lysine decarboxylase enzyme activity, takes charge of the diamine and polyamine formations (**Figure 3**) [45, 48, 49].

Spermidine synthase catalyzes spermidine formation from putrescine (**Figure 3**) [50, 51]. Spermidine is a precursor of other polyamines such as spermine and structural isomer thermospermine [45, 52]. Spermidine, regulating several crucial biological processes (Na^+-K^+ ATPaz), protects the membrane potential and controls the intracellular pH and volume [53]. Besides, spermidine, a polyamine found in the cellular metabolism, has a role in the neuronal nitric oxide synthase inhibitions and intestinal tissue developments [54].

Spermine, whose precursor amino acid is ornithine, is formed from spermidine through the spermine synthase enzyme (**Figure 3**) [51]. Spermine is present in several organisms and tissues, as it is a polyamine that is found in all eukaryotic cells and has a role in the cellular metabolism [52, 55]. It plays a role in the intestinal tissue developments and stabilizes the helical structure in viruses [52, 56, 57].

Agmatine is a biogenic amine formed by arginine decarboxylase enzyme activity of arginine amino acid (**Figure 3**) [12, 44, 58]. Agmatine participates in the polyamine metabolism over the putrescine hydrolyzed by the agmatine enzyme and has several functions such as nitric oxide synthesis regulation, polyamine metabolism, and matrix metalloproteinase and enzyme activity leading to H_2O_2 production [59, 60].

The detoxification system, splitting the biogenic amines in the human body, consists of monoamine oxidase (MAO), diamine oxidase (DAO), polyamine oxidase (PAO), and histamine-N-methyl transferase (HNMT) [17, 61, 62].

3. Effect of biogenic amines on health

Biogenic amines have several important biological roles in the body and constitute the first step of protein, hormone, and nucleic acid synthesis [61, 63]. The polyamines such as putrescine, spermine, and spermidine are the unique components of living cells. Besides, the polyamines were stated to require maintaining the intestinal immunologic systems and healthy metabolic function activities [52, 64–67]. The biogenic amines cause respiratory disorders, headache, tachycardia, hypo- or hypertension, and allergic reactions when taken in high concentrations together with nutrients [68].

Biogenic amines are vasoactive components, and taking them in high amounts leads to change in blood pressure in humans and animals. The amines bear essential psychoactive or vasoactive effects, as they have the biological activities such as histamine, tryptamine, tyramine, and phenylethylamine [33]. Histamine is a biologically active amine and quickly scatters to the tissues through blood circulation and leads to several reactions. However, in the case where aminoxidase enzyme inhibitors are present in the environment, the biogenic amine prevents detoxification, and health problems (erythema, edema, rash, headache, burning, etc.) show up [17, 68]. Histamine also has essential metabolic functions such as a role in the nervous system functions and blood pressure control. It mainly takes effect by binding to the cardiovascular system (vasodilatation and hypotension) and cell membrane receptors in several secretory glands (such as gastric acid secretion) [22, 23]. In addition to them, it may lead to some neurotransmission disorders and causes headache, flushing, gastrointestinal disorders, and edema by giving rise to blood vessel dilatations [61, 69]. Histamine intoxicates when orally taken in amounts of 8 mg and above [3]. Individuals generally have lower intestinal oxidase enzyme activities according to the healthy persons, as they hold the gastrointestinal problems such as gastritis, stomach and colonic ulcers [6, 69].

Intestinal mucosal injuries may decrease the enzyme functions by detoxifying the biogenic amines [17, 63]. The DAO activity disruption causes histamine intolerance and also allergic reactions as a result of the drug utilization, as it is caused by genetic and gastrointestinal diseases or DAO inhibition [17, 70]. It was found to increase the histamine toxicity by preventing the histamine oxidation of putrescine, cadaverine, and agmatine in humans [71].

Biogenic amines lead to hypertension, as they have vasoconstriction effects such as tyramine, phenylethylamine, and tryptamine [37, 68, 72]. Consuming tyramine-rich nutriments was found to react with the tyramine MAO inhibitor drugs and cause hypertensive crisis and also migraine in some patients [73]. Tyramine is revealed to inhibit MAO, tryptamine DAO, phenylethylamine DAO, and HNMT enzymes [74, 75].

In the case of deficiency of putrescine, found in the high concentration in brains, is stated to develop the depression and also useful in the depression physiopathology [3, 76].

The pharmacological effects of putrescine, cadaverine, spermine, and spermidine are at lower levels according to histamine, tyramine, and phenylethylamine [77]. Putrescine, causing hypotension, bradycardia, and lockjaw, creates carcinogenic heterocyclic compounds including nitrosamine, nitrosopyrrolidine, and nitrosopiperidine as some biogenic amines such as cadaverine, spermine, and spermidine react with the nitrite [5, 10, 39, 78].

Polyamines are known to lead to low-dose colon cancer by affecting the cell developments and differentiation [79, 80]. In addition to them, putrescine, cadaverine, spermine, and spermidine were also found to induce apoptosis and inhibit cell proliferation. The high-dose putrescine was found to induce apoptosis and prevent the spread [81, 82]. This putrescine effect pertains to increasing the nitric oxide synthesis, inhibiting the redox reactions and binding directly to the carcinogenic agents [82].

Eating disorders such as anorexia nervosa and bulimia nervosa disrupt the function of brain serotonin [83]. Albumin deficiency shows up due to inadequate nutrition, and tryptophan transition from the blood-brain barrier increases, as it could not connect to the albumin. As a consequence, an increase occurs in the brain serotonin concentration [84, 85]. The drugs (MAO inhibitors) are used, as they change the serotonin levels in the depression, generalized anxiety disorder, and social phobia treatments [86]. The MAO inhibitors, used in treating these diseases, increase the brain concentrations by preventing the neurotransmitter (serotonin) disruptions [73, 86].

Agmatine shows a nephroprotective effect by increasing the glomerular filtration rate, and it also has a hypoglycemic impact as a result of several molecular mechanisms taking place in the blood glucose regulation [56, 87]. Besides, the agmatine level of schizophrenia patients was shown to be higher compared to that of healthy humans [88].

4. Conclusion

The present information related to biogenic amines having different physiological functions and similar chemical structures and metabolic pathways was updated, undesirable effects were considered more comprehensively for human and animal health, and information was submitted about the essential diseases caused by biogenic amines.

Author details

Dincer Erdag[1], Oguz Merhan[2*] and Baris Yildiz[3]

1 Department of Medical Services and Techniques, Atatürk Vocational School of Health Services, Kafkas University, Kars, Turkey

2 Department of Biochemistry, Faculty of Veterinary, Kafkas University, Kars, Turkey

3 Department of Physiology, Faculty of Veterinary, Kafkas University, Kars, Turkey

*Address all correspondence to: oguzmerhan@hotmail.com

IntechOpen

References

[1] Cellini B, Montioli R, Oppici E, Astegno A, Voltattorni CB. The chaperone role of the pyridoxal 5′-phosphate and its implications for rare diseases involving B6-dependent enzymes. Clinical Biochemistry. 2014;**47**:158-165. DOI: 10.1016/j.clinbiochem.2013.11.021

[2] Liu X, Yang LX, Lu YT. Determination of biogenic amines by 3-(2-furoyl)quinoline-2carboxaldehyde and capillary electrophoresis with laser-induced fluorescence detection. Journal of Chromatography A. 2003;**998**:213-219. DOI: 10.1016/S0021-9673(03)00637-X

[3] Doğan A, Otlu S, Büyük F, Aksu P, Tazegül E, Erdağ D. Effects of cysteamine, putrescine and cysteamine-putrescine combination on some bacterium. Kafkas Universitesi Veteriner Fakultesi Dergisi. 2012;**18**:1015-1019. DOI: 10.9775/kvfd.2012.6928

[4] Tiqua SM. Metabolic diversity of the heterotrophic microorganisms and potential link to pollution of the Rouge River. Environmental Pollution. 2010;**158**:1435-1443. DOI: 10.1016/j.envpol.2009.12.035

[5] Biji KB, Ravishankar CN, Venkateswarlu R, Mohan CO, Gopal TKS. Biogenic amines in seafood: A review. Journal of Food Science and Technology. 2016;**53**:2210-2218. DOI: 10.1007/s13197-016-2224-x

[6] Jairath G, Singh PK, Dabur RS, Rani M, Chaudhari M. Biogenic amines in meat and meat products and its public health significance: A review. Journal of Food Science and Technology. 2015;**52**:6835-6846. DOI: 10.1007/s13197-015-1860-x

[7] Mafra I, Herbert P, Santos L, Barros P, Alves A. Evaluation of biogenic amines in some Portuguese quality wines by HPLC fluorescence detection of OPA derivatives. American Journal of Enology and Viticulture. 1999;**50**:128-132

[8] Düz M, Fidan AF. Biogenic amines and effects. Kocatepe Veterinary Journal. 2016;**9**:114-121. DOI: 10.5578/kvj.23113

[9] Spano G, Russo P, Lonvaud-Funel A, Lucas P, Alexandre H, Grandvalet C, et al. Biogenic amines in fermented foods. European Journal of Clinical Nutrition. 2010;**64**:95-100

[10] Shalaby AR. Significance of biogenic amines to food safety and human health. Food Research International. 1996;**29**:675-690. DOI: 10.1016/S0963-9969(96)00066-X

[11] Hofto LR, Lee CE, Cafiero M. The importance of aromatic-type interactions in serotonin synthesis: Protein-ligand interactions in tryptophan hydroxylase and aromatic amino acid decarboxylase. Journal of Computational Chemistry. 2009;**30**:1111-1115. DOI: 10.1002/jcc.21139

[12] Landete JM, Arena ME, Pardo I, Manca de Nadra MC, Ferrer S. The role of two families of bacterial enzymes in putrescine synthesis from agmatine via agmatine deiminase. International Microbiology. 2010;**13**:169-177. DOI: 10.2436/20.1501.01.123

[13] Pessione E, Cirrincione S. Bioactive molecules released in food by lactic acid bacteria: Encrypted peptides and biogenic amines. Frontiers in Microbiology. 2016;**7**:1-19. DOI: 10.3389/fmicb.2016.00876

[14] Bjelakovic G, Stojanovic I, Stoimenov TJ, Pavlovic D, Kocic G,

Bjelakovic GB, et al. Polyamines, folic acid supplementation and cancerogenesis. Pteridines. 2017;**28**:115-131. DOI: 10.1515/pterid-2017-0012

[15] Soleas GJ, Carey M, Goldberg DM. Method development and cultivar-related differences of nine biogenic amines in Ontario wines. Food Chemistry. 1999;**64**:49-58. DOI: 10.1016/S0308-8146(98)00092-2

[16] Moya-Garcia AA, Pino-Angeles A, Gil-Redondo R, Morreale A, Sanchez-Jimenez F. Structural features of mammalian histidine decarboxylase reveal the basis for specific inhibition. British Journal of Pharmacology. 2009;**157**:4-13. DOI: 10.1111/j.1476-5381.2009.00219.x

[17] Maintz L, Novak N. Histamine and histamine intolerance. The American Journal of Clinical Nutrition. 2007;**85**:1185-1196. DOI: 10.1093/ajcn/85.5.1185

[18] Criado PR, Jardim Criado RF, Maruta CW, Machado Filho CA. Histamine, histamine receptors and antihistamines: New concepts. Anais Brasileiros de Dermatologia. 2010;**85**:195-210

[19] Jeong CD, Mamuad LL, Kim SH, Choi YJ, Soriano AP, Cho KK, et al. Effect of soybean meal and soluble starch on biogenic amine production and microbial diversity using in vitro rumen fermentation. Asian-Australasian Journal of Animal Sciences. 2015;**28**:50-57. DOI: 10.5713/ajas.14.0555

[20] Shahid M, Tripathi T, Sobia F, Moin S, Siddiqui M, Khan RA. Histamine, histamine receptors, and their role in immunomodulation: An updated systematic review. The Open Immunology Journal. 2009;**2**:9-41. DOI: 10.2174/1874226200902010009

[21] Qin L, Zhao D, Xu J, Ren X, Terwilliger EF, Parangi S, et al. The vascular permeabilizing factors histamine and serotonin induce angiogenesis through TR3/Nur77 and subsequently truncate it through thrombospondin-1. Blood. 2013;**121**:2154-2164. DOI: 10.1182/blood-2012-07-443903

[22] Triggiani M, Patella V, Staiano RI, Granata F, Marone G. Allergy and the cardiovascular system. Clinical and Experimental Immunology. 2008;**153**:7-11. DOI: 10.1111/j.1365-2249.2008.03714.x

[23] Nakamura Y, Ishimaru K, Shibata S, Nakao A. Regulation of plasma histamine levels by the mast cell clock and its modulation by stress. Scientific Reports. 2017;**7**:1-12. DOI: 10.1038/srep39934

[24] Yuwen L, Zhang FL, Chen QH, Lin SJ, Zhao YL, Li ZY. The role of aromatic L-amino acid decarboxylase in bacillamide C biosynthesis by *Bacillus atrophaeus* C89. Scientific Reports. 2013;**3**:1-10. DOI: 10.1038/srep01753

[25] Torrens-Spence MP, Liu P, Ding H, Harich K, Gillaspy G, Li J. Biochemical evaluation of the decarboxylation and decarboxylation-deamination activities of plant aromatic amino acid decarboxylases. The Journal of Biological Chemistry. 2013;**288**:2376-2387. DOI: 10.1074/jbc.M112.401752

[26] Tittarelli R, Mannocchi G, Pantano F, Romolo FS. Recreational use, analysis and toxicity of tryptamines. Current Neuropharmacology. 2015;**13**:26-46. DOI: 10.2174/1570159X1366614121022 2409

[27] Önal A. A review: Current analytical methods for the determination of biogenic amines in foods. Food Chemistry. 2007;**103**:1475-1486. DOI: 10.1016/j.foodchem.2006.08.028

[28] Berry MD. Mammalian central nervous system trace amines. Pharmacologic amphetamines, physiologic neuromodulators. Journal of Neurochemistry. 2004;**90**:257-271. DOI: 10.1111/j.1471-4159.2004.02501.x

[29] Tieman D, Taylor M, Schauer N, Fernie AR, Hanson AD, Klee HJ. Tomato aromatic amino acid decarboxylases participate in synthesis of the flavor volatiles 2-phenylethanol and 2-phenylacetaldehyde. Proceedings of the National Academy of Sciences of the United States of America. 2006;**103**:8287-8292. DOI: 10.1073/pnas.0602469103

[30] Parthasarathy A, Cross PJ, Dobson RCJ, Adams LE, Savka MA, Hudson AO. A three-ring circus: Metabolism of the three proteogenic aromatic amino acids and their role in the health of plants and animals. Frontiers in Molecular Biosciences. 2018;**5**:1-30. DOI: 10.3389/fmolb.2018.00029

[31] Irsfeld M, Spadafore M, Prüß BM. β-Phenylethylamine, a small molecule with a large impact. Webmedcentral. 2013;**4**:1-15

[32] Kaur N, Kumari B. Phenylethylamıne: Health benefits—A review. World Journal of Pharmacy and Pharmaceutical Sciences. 2016;**5**:743-750. DOI: 10.20959/wjpps20164-6429

[33] Connil N, Le Breton Y, Dousset X, Auffray Y, Rince A, Prevost H. Identification of the *Enterococcus faecalis* tyrosine decarboxylase operon involved in tyramine production. Applied and Environmental Microbiology. 2002;**68**:3537-3544. DOI: 10.1128/AEM.68.7.3537-3544.2002

[34] Bargossi E, Tabanelli G, Montanari C, Lanciotti R, Gatto V, Gardini F, et al. Tyrosine decarboxylase activity of enterococci grown in media with different nutritional potential: Tyramine

and 2-phenylethylamine accumulation and *tyrDC* gene expression. Frontiers in Microbiology. 2015;**6**:1-10. DOI: 10.3389/fmicb.2015.00259

[35] Zhu H, Xu G, Zhang K, Kong X, Han R, Zhou J, et al. Crystal structure of tyrosine decarboxylase and identification of key residues involved in conformational swing and substrate binding. Scientific Reports. 2016;**6**:1-10. DOI: 10.1038/srep27779

[36] Gonzalez-Jimenez M, Arenas-Valganon J, Garcia-Santos MP, Calle E, Casado J. Mutagenic products are promoted in the nitrosation of tyramine. Food Chemistry. 2017;**216**:60-65. DOI: 10.1016/j.foodchem.2016.08.006

[37] Anwar MA, Ford WR, Broadley KJ, Herbert AA. Vasoconstrictor and vasodilator responses to tryptamine of rat-isolated perfused mesentery: Comparison with tyramine and b-phenylethylamine. British Journal of Pharmacology. 2012;**165**:2191-2202. DOI: 10.1111/j.1476-5381.2011.01706.x

[38] Jacob G, Gamboa A, Diedrich A, Shibao C, Robertson D, Biaggioni I. Tyramine-induced vasodilation mediated by dopamine contamination: A paradox resolved. Hypertension. 2005;**46**:355-359. DOI: 10.1161/01.HYP.0000172353.62657.8b

[39] Benkerroum N. Biogenic amines in dairy products: Origin, incidence, and control means. Comprehensive Reviews in Food Science and Food Safety. 2016;**15**:801-826. DOI: 10.1111/1541-4337.12212

[40] Cote F, Thevenot E, Fligny C, Fromes Y, Darmon M, Ripoche MA, et al. Disruption of the nonneuronal tph1 gene demonstrates the importance of peripheral serotonin in cardiac function. Proceedings of the National Academy of Sciences of the United States of America. 2003;**100**:13525-13530. DOI: 10.1073/pnas.2233056100

[41] Perez-Garcia G, Meneses A. Memory formation, amnesia, improved memory and reversed amnesia: 5-HT role. Behavioural Brain Research. 2008;**195**:17-29. DOI: 10.1016/j. bbr.2007.11.027

[42] Zhang X, Yan H, Luo Y, Huang Z, Rao Y. Thermoregulation-independent regulation of sleep by serotonin revealed in mice defective in serotonin synthesis. Molecular Pharmacology Fast Forward. 2018;**93**:657-664. DOI: 10.1124/mol.117.111229

[43] Nakada Y, Itoh Y. Identification of the putrescine biosynthetic genes in *Pseudomonas aeruginosa* and characterization of agmatine deiminase and N-carbamoylputrescine amidohydrolase of the arginine decarboxylase pathway. Microbiology. 2003;**149**:707-714. DOI: 10.1099/mic.0.26009-0

[44] Hao YJ, Kitashiba H, Honda C, Nada K, Moriguchi T. Expression of arginine decarboxylase and ornithine decarboxylase genes in apple cells and stressed shoots. Journal of Experimental Botany. 2005;**56**:1105-1115. DOI: 10.1093/jxb/eri102

[45] Pegg AE, Casero RA. Current status of the polyamine research field. Methods in Molecular Biology. 2011;**720**:3-35. DOI: 10.1007/978-1-61779-034-8_1

[46] Valdes Santiago L, Ruiz Herrera J. Stress and polyamine metabolism in fungi. Frontiers in Chemistry. 2013;**1**: 1-10. DOI: 10.3389/fchem.2013.00042

[47] Ioannidis NE, Sfichi L, Kotzabasis K. Putrescine stimulates chemiosmotic ATP synthesis. Biochimica et Biophysica Acta. 2006;**1757**:821-828. DOI: 10.1016/j. bbabio.2006.05.034

[48] Jeong S, Yeon YJ, Choi EG, Byun S, Cho DH, Kim Il K, et al. Alkaliphilic lysine decarboxylases for effective synthesis of cadaverine from L-lysine. Korean Journal of Chemical Engineering. 2016;**33**:1530-1533. DOI: 10.1007/s11814-016-0079-5

[49] Sagong HY, Kim KJ. Lysine decarboxylase with an enhanced affinity for pyridoxal 5-phosphate by disulfide bond-mediated spatial reconstitution. PLoS One. 2017;**12**:e0170163. DOI: 10.1371/journal.pone.0170163

[50] Wu H, Min J, Ikeguchi Y, Zeng H, Dong A, Loppnau P, et al. Structure and mechanism of spermidine synthases. Biochemistry. 2007;**46**:8331-8339. DOI: 10.1021/bi602498k

[51] Sanchez-Jimenez F, Ruiz-Perez MV, Urdiales JL, Medina MA. Pharmacological potential of biogenic amine–polyamine interactions beyond neurotransmission. British Journal of Pharmacology. 2013;**170**:4-16. DOI: 10.1111/bph.12109

[52] Takahashi T, Kakehi JI. Polyamines: Ubiquitous polycations with unique roles in growth and stress responses. Annals of Botany. 2010;**105**:1-6. DOI: 10.1093/aob/mcp259

[53] Carvalho FB, Mello CF, Marisco PC, Tonello R, Girardi BA, Ferreira J, et al. Spermidine decreases Na$^+$, K$^+$-ATPase activity through NMDA receptor and protein kinase G activation in the hippocampus of rats. European Journal of Pharmacology. 2012;**684**:79-86. DOI: 10.1016/j.ejphar.2012.03.046

[54] Hu J, Mahmoud MI, E1-Fakahany EE. Polyamines inhibit nitric oxide synthase in rat cerebellum. Neuroscience Letters. 1994;**175**:41-45

[55] Paschalidis KA, Roubelakis-Angelakis KA. Spatial and temporal distribution of polyamine levels and polyamine anabolism in differento gans/tissues of the tobacco plant. Correlations with age, cell division/expansion, and differentiation. Plant Physiology.

2005;**138**:142-152. DOI: 10.1104/pp.104.055483

[56] Medina MA, Urdiales JL, Rodriguez-Caso C, Ramirez FJ, Sanchez-Jimenez F. Biogenic amines and polyamines: Similar biochemistry for different physiological missions and biomedical applications. Critical Reviews in Biochemistry and Molecular Biology. 2003;**38**:23-59. DOI: 10.1080/713609209

[57] Welsh PA, Kuhn SS, Prakashagowda C, McCloskey D, Feith D. Spermine synthase overexpression in vivo does not increase susceptibility to DMBA/TPA skin carcinogenesis or Min-Apc intestinal tumorigenesis. Cancer Biology & Therapy. 2012;**13**:358-368. DOI: 10.4161/cbt.19241

[58] Morris SM. Arginine: Beyond protein. The American Journal of Clinical Nutrition. 2006;**83**:508-512. DOI: 10.1093/ajcn/83.2.508S

[59] Demady DR, Jianmongkol S, Vuletich JL, Bender AT, Osawa Y. Agmatine enhances the NADPH oxidase activity of neuronal NO synthase and leads to oxidative inactivation of the enzyme. Molecular Pharmacology. 2001;**59**:24-29

[60] Pegg AE. Mammalian polyamine metabolism and function. International Union of Biochemistry and Molecular Biology Life. 2009;**61**:880-894. DOI: 10.1002/iub.230

[61] Santos MHS. Biogenic amines: Their importance in foods. International Journal of Food Microbiology. 1996;**29**:213-231

[62] Schwelberger HG, Feurle J, Houen G. Mapping of the binding sites of human histamine N-methyltransferase (HNMT) monoclonal antibodies. Inflammation Research. 2017;**66**:1021-1029. DOI: 10.1007/s00011-017-1086-7

[63] Karovicova J, Kohajdova Z. Biogenic amines in food. Chemical Papers. 2005;**59**:70-79

[64] Yatin M. Polyamines in living organisms. Journal of Cell and Molecular Biology. 2002;**1**:57-67

[65] Soda K. The mechanisms by which polyamines accelerate tumor spread. Journal of Experimental & Clinical Cancer Research. 2011;**30**:1-9. DOI: 10.1186/1756-9966-30-95

[66] Mandal S, Mandal A, Johansson HE, Orjalo AV, Park MH. Depletion of cellular polyamines, spermidine and spermine, causes a total arrest in translation and growth in mammalian cells. Proceedings of The National Academy of Sciences of The United States of America. 2013;**110**:2169-2174. DOI: 10.1073/pnas.1219002110

[67] Pegg AE. Functions of polyamines in mammals. The Journal of Biological Chemistry. 2016;**291**:14904-14912. DOI: 10.1074/jbc.R116.731661

[68] Ladero V, Calles M, Fernandez M, Alvarez MA. Toxicological effects of dietary biogenic amines. Current Nutrition and Food Science. 2010;**6**:145-156. DOI: 10.2174/157340110791233256

[69] Stadnik J, Dolatowski ZJ. Biogenic amines in meat and fermented meat products. Acta Scientiarum Polonorum Technologia Alimentaria. 2010;**9**:251-263

[70] Kovacova-Hanuskova E, Buday T, Gavliakova S, Plevkova J. Histamine, histamine intoxication and intolerance. Allergologia et Immunopathologia. 2015;**43**:498-506. DOI: 10.1016/j.aller.2015.05.001

[71] Gardini F, Özogul Y, Suzzi G, Tabanelli G, Özogul F. Technological factors affecting biogenic amine content in foods: A review. Frontiers

in Microbiology. 2016;7:1-18. DOI: 10.3389/fmicb.2016.01218

[72] Anwar MA, Ford WR, Herbert AA, Broadley KJ. Signal transduction and modulating pathways in tryptamine-evoked vasopressor responses of the rat isolated perfused mesenteric bed. Vascular Pharmacology. 2013;**58**: 140-149. DOI: 10.1016/j.vph.2012.10.007

[73] Fisar Z. Drugs related to monoamine oxidase activity. Progress in Neuro-Psychopharmacology & Biological Psychiatry. 2016;**69**:112-124. DOI: 10.1016/j.pnpbp.2016.02.012

[74] Ordonez AI, Ibanez FC, Torre P, Barcina Y. Formation of biogenic amines in Idiazabal ewe's-milk cheese: Effect of ripening, pasteurization, and starter. Journal of Food Protection. 1997;**60**:1371-1375

[75] San Mauro Martin I, Brachero S, Garicano Vilar E. Histamine intolerance and dietary management: A complete review. Allergologia et Immunopathologia. 2016;**44**:475-483. DOI: 10.1016/j.aller.2016.04.015

[76] Zomkowski ADE, Santos ARS, Rodrigues ALS. Putrescine produces antidepressant-like effects in the forced swimming test and in the tail suspension test in mice. Progress in Neuro-Psychopharmacology & Biological Psychiatry. 2006;**30**:1419-1425. DOI: 10.1016/j.pnpbp.2006.05.016

[77] Halasz A, Barath A, Simon-Sarkadi L, Holzapfel W. Biogenic amines and their production by microorganisms in food. Trends in Food Science & Technology. 1994;**5**:42-49. DOI: 10.1016/0924-2244(94)90070-1

[78] Atakisi E, Merhan O. Nitric oxide synthase and nitric oxide involvement in different toxicities. In: Saravi SSS, editor. Nitric Oxide Synthase Simple Enzyme-Complex Roles. 1st ed. Croatia:

InTech; 2017. p. 197-214. DOI: 10.5772/intechopen.68494

[79] Milovica V, Turchanowa L, Khomutov AR, Khomutov RM, Caspary WF, Stein J. Hydroxylamine containing inhibitors of polyamine biosynthesis and impairment of colon cancercell growth. Biochemical Pharmacology. 2001;**61**:199-206

[80] Linsalata M, Notarnicola M, Tutino V, Bifulco M, Santoro A, Laezza C, et al. Effects of anandamide on polyamine levels and cell growth in human colon cancer cells. Anticancer Research. 2010;**30**:2583-2589

[81] Takao K, Rickhag M, Hegardt C, Oredsson S, Persson L. Induction of apoptotic cell death by putrescine. The International Journal of Biochemistry & Cell Biology. 2006;**38**:621-628. DOI: 10.1016/j.biocel.2005.10.020

[82] Doğan A, Aksu Kılıçle P, Erdağ D, Doğan ANC, Özcan K, Doğan E. The protective effects of cysteamine, putrescine, and the combination of cysteamine and putrescine on fibrosarcoma induced in mice with 3-methylcholanthrene. Turkish Journal of Veterinary and Animal Sciences. 2016;**40**:575-582. DOI: 10.3906/vet-1510-49

[83] Kaye W. Neurobiology of anorexia and bulimia nervosa. Physiology & Behavior. 2008;**94**:121-135. DOI: 10.1016/j.physbeh.2007.11.037

[84] Richard DM, Dawes MA, Mathias CW, Acheson A, Hill-Kapturczak N, Dougherty DM. L-Tryptophan: Basic metabolic functions, behavioral research and therapeutic indications. International Journal of Tryptophan Research. 2009;**2**:45-60

[85] Jenkins TA, Nguyen JCD, Polglaze KE, Bertrand PP. Influence of tryptophan and serotonin on mood and

cognition with a possible role of the gut-brain axis. Nutrients. 2016;**8**:1-15. DOI: 10.3390/nu8010056

[86] Riederer P, Laux G. MAO-inhibitors in Parkinson's disease. Experimental Neurobiology. 2011;**20**:1-17. DOI: 10.5607/en.2011.20.1.1

[87] Satriano J. Arginine pathways and the inflammatory response: Interregulation of nitric oxide and polyamines: Review article. Amino Acids. 2004;**26**:321-329. DOI: 10.1007/s00726-004-0078-4

[88] Liu P, Jing Y, Collie ND, Dean B, Bilkey DK, Zhang H. Altered brain arginine metabolism in schizophrenia. Translational Psychiatry. 2016;**6**:1-9. DOI: 10.1038/tp.2016.144